Inspiring Science in the Early Years

Inspiring Science in the Early Years

Exploring Good Practice

Edited by Di Stead and Lois Kelly

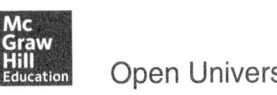 Open University Press

Open University Press
McGraw-Hill Education
McGraw-Hill House
Shoppenhangers Road
Maidenhead
Berkshire
England
SL6 2QL

email: enquiries@openup.co.uk
world wide web: www.openup.co.uk

and Two Penn Plaza, New York, NY 10121-2289, USA

First published 2015

A catalogue record of this book is available from the British Library

ISBN-13: 978-0-33-526452-0 (pb)
ISBN-10: 0-33-526452-2 (pb)
eISBN: 978-0-33-526453-7

Library of Congress Cataloging-in-Publication Data
CIP data applied for

Typesetting and e-book compilations by
RefineCatch Limited, Bungay, Suffolk

Fictitious names of companies, products, people, characters and/or data that may be used herein (in case studies or in examples) are not intended to represent any real individual, company, product or event.

Printed and bound by CPI Group (UK) Ltd, Croydon, CR0 4YY

Praise for this book

"This clearly written and engaging book examines Science in the Early Years through a variety of activities, including role-play, toys and technology. The vital importance of sensory experiences and language is emphasized throughout.

The wide experience and knowledge of the authors guarantees a highly enjoyable read. The links to all curricula in the UK are extremely beneficial and I particularly liked the way that photographs and Key Points text boxes have been used throughout the book. The breadth and depth of writing about science makes this a highly desirable book for any practitioner working or studying in the Early Years."

Kathy Brodie, Independent Early Years Consultant

"As an Early Years consultant who is passionate about children's thinking, exploring, questioning, investigating and most of all engaging…I really enjoyed this book. I especially liked it because it provokes practitioners to think about 'science' as the discovery and exploration of the world around us and not just as a National Curriculum subject. The mix of authors, their writing styles and the content of each chapter makes it a really easy and engaging read. Definitely one to add to your reading list if you work with children in the Early Years."

Alistair Bryce-Clegg, Early Years Consultant

"As the title suggests, this book from the first page onwards inspires the reader to learn more about how to develop, enhance and incorporate effective practice in science in the early years. In addition to developing an understanding of how to approach the teaching of science, it gives a clearly articulated and accessible theoretical insight into how young children learn. To compliment this there are points of reflection, case studies, practical tasks and examples from the field. This is a valuable book for both students and practitioners alike as it goes beyond just giving suggestions for what to do; it explains the why and the how as well."

Joanne McNulty, Manchester Metropolitan University

"This is a warm, accessible book, strongly grounded in research. It interweaves real life examples of science in the early years with underlying pedagogic principles and inspires new possibilities. The enthusiasm of the authors is contagious!"

Kendra McMahon, Bath Spa University

Contents

List of figures and tables

Figures

Tables

Notes on contributors

The editors

The editors have extensive experience of teaching in primary schools and as teacher educators. Including science in the early years curriculum has interested them for a number of years. Lois Kelly and Di Stead are co-editors of *Enhancing Primary Science: Developing Cross-Curricular Links*, published by Open University Press in late 2012.

Lois Kelly is a teacher educator and education consultant and has international experience. She worked in teacher education for over 10 years, where she was curriculum coordinator for primary science on BA(QTS) and PGCE courses and developed science modules for students on early years courses. She has worked at Liverpool Hope University, Manchester Metropolitan University and University of Chester. Before that she taught across the primary age range. She contributed to C. Bold (ed.) (2011) *Supporting Learning and Teaching* (London: Routledge). She is a member of the Primary Committee of the Association for Science Education (ASE) and has presented workshops at ASE. She is a hub leader for the Primary Science Quality Mark.

Di Stead is an educational consultant, working with primary teachers and their schools. She has provided science in-service training from the north of England to India. She worked for two decades in higher education, at Liverpool Hope University for over 17 years and before that at Bishop Grosseteste College in Lincoln and latterly at University of Chester. She learned the importance of including science in an early years curriculum through providing support for students on placement in early years settings and teaching on both the undergraduate and postgraduate initial teacher training courses. She learned her craft teaching in a primary school in the east end of Sheffield for 14 years.

The authors

This book draws on current practice of the following experts in the field of early years education.

Babs Anderson is Lecturer in Childhood Studies at Liverpool Hope University. Her first degree was in environmental biology and she has maintained an interest in science and mathematics. She taught for 25 years, most of these part-time in a primary school, focusing on the early years. In addition, she was a freelance consultant for Knowsley Healthy Schools, advising and supporting schools to gain accredited Healthy School status. One of her consultant programmes was 'Listening to Young Children', working with early years practitioners in children's centres in two local authorities in the north-west, supporting them in developing effective communication. Her research interests include young children's collaborative learning, including when playing in continuous provision, the use of language as a cultural tool, young children's thinking skills, shared sustained thinking and the co-construction of knowledge and understanding, within adult–child dyads, child–child dyads and collaborative groups.

Linda Atherton is an independent consultant, who has previously worked for over 10 years as the science adviser for Warwickshire. A chartered science graduate with over 25 years' teaching experience in both primary and secondary phases, Linda has a wide and varied background of teaching and leading all areas of science in schools. Linda has vast knowledge of working closely with primary teachers and teams to develop their skills and share strategies, methods and materials. Her multi-sensory approach is enhanced by making strong cross-curricular links with literacy, numeracy, humanities, design and technology, drama and art. Her groundbreaking work on creative science and cross-curricular work has been recognized by The Wellcome Trust. A published author (see L. Atherton and N. Cronin (2007) *Focus Weeks: Science* (Key Stage 1) (Leamington Spa: Language Centre Publications) as an example of her work), Linda has also written and delivered highly acclaimed science courses and workshops both regionally and nationally for local authorities, independent schools, science learning centres and the Association of Science Education (ASE).

Jessica Baines Holmes is Senior Lecturer within the School of Education at the University of Brighton specializing in teaching primary science. She teaches on both the BA (QTS) 3–7 and PGCE 3–7 routes. Prior to this she taught for over 10 years in primary schools in London and Sussex, primarily in the foundation stage and Key Stage 1, and has been a subject leader for science.

Faith Fletcher is a senior lecturer at the University of Chester, where she teaches on both the postgraduate and undergraduate early years programmes. She has enjoyed a long career working with children from the ages of 3 to 18 years old in a variety of settings and subject areas. Her experience with older children and special education drew her to the early years and the notion of quality early experiences and intervention. The last 12 years of teaching were in the Foundation Stage, while working towards a MA in Education with early years focus at Manchester Metropolitan University. At Liverpool Hope University she was a member of the Early Childhood Studies team and also worked with the Professional Studies team on the QTS course and contributed to PGCE programme. Her interest in developing high-quality reflective practice in the early years sector led to her involvement in the EYPS programme based at Hope but combining the expertise of three universities in the north-west. Faith's research interests lie in the area of children's rights and the voice of the child, with a particular emphasis on the adult's understanding of their role in supporting and developing these rights.

Eleanor Hoskins is Senior Lecturer within the Faculty of Education at Manchester Metropolitan University, where she specializes in teaching early years and primary science. Within this role, she has been responsible for introducing and teaching new early years science sessions within the postgraduate and undergraduate initial teacher training programmes and supervises many trainee teachers throughout their school placements. Before this, she taught for 10 years in several primary and early years settings across two different local authorities and gained experience as a subject leader in both science and mathematics, as well as taking responsibilities for assessment and as SENCo. In addition, she has also gained experiences as a school improvement teacher for Manchester local authority, assistant head and deputy headteacher. Within her management role in school, she contributed to the creation and organization of a new, functioning, open-plan foundation stage before other schools and settings had progressed to this level. Further expertise within the early years foundation stage involved trialling new approaches to continuous provision learning and taking a proactive lead with children, parents and staff to ensure thorough transition for children between the early years foundation stage and Key Stage 1. Her research interests centre around teaching early years science alongside technology and her master's degree dissertation examined young children's capabilities with the use of technology to support their early scientific explorations.

Kathleen Orlandi has been a senior lecturer at Liverpool Hope University for 10 years. She developed a PGCE in Foundation Stage and Key Stage 1 and now leads the Childhood Studies team. She is a Senior Fellow of the Higher Education Academy. Kathleen was a teacher of biology in a secondary school in London for 10 years at the start of her teaching career. Later she worked in the primary sector and developed an interest in very young children and their learning. She was the coordinator of the Foundation Stage when the first Foundation Stage guidance was published in 2000. Her doctoral research was about the impact of policy on practice in the foundation stage. It examined the nature of children's experiences in the light of the culture and policy that drives provision. This research informed her 2012 book *Onwards and Upwards: Supporting Transition to Key Stage One* (Abingdon: Routledge). She has a keen interest in learning and teaching theory both for young children and for the adults who work with them. She believes that professionals working with children need continuing professional development that allows them to develop expertise through research and enquiry, rather than training. This is explored in her article, written with Babs Anderson, 'Researching the Role of Dialogue, Writing and Critical Reflection in Unlearning for Students with Professional Backgrounds', published in the online journal *Teaching Anthropology* in 2012.

Kathy Schofield is currently College Director for the Primary Science Teaching Trust (PSTT). Previously she was a senior lecturer in primary science at Manchester Metropolitan University (MMU), where she was also contracted by the Science Learning Centre North West to deliver continuing professional development. During her time at MMU she achieved a STEM masters, researching collaborative teaching and learning and exploring creativity in developing learning beyond the classroom. She also has over 10 years' experience as a primary teacher and science subject leader across all phases. Kathy's current role for PSTT allows her to indulge her passion for promoting and inspiring teachers to deliver practical primary science from early years foundation stage to Key Stage 3.

Acknowledgements

We would like to thank the following for their contributions to this book.

The staff and children at Anfield Infant and Early Years School, Liverpool, in particular Katie Louise Ledgerton.

The parents and children at *St Gabriel's Praise and Play*, Stockport.

Mrs Mavis Smith, Headteacher at Rainhill Community Nursery, Merseyside.

Illustration credits:

Figures 1.1, 1.2, 2.1, 2.2, 2.3, 2.4, 3.1, 3.2, 3.3, 3.4, 3.5, 4.1, 4.2, 4.3, 4.4, 5.1, 5.2, 5.3, 5.4, 5.5, 5.6, 5.7, 6.1, 6.2, 6.3, 7.1, 7.2, 7.3, 7.4, 7.5, 7.6, 7.7, 7.8, 8.1, 8.2, 9.1, 9.3, reproduced with thanks to Di Stead and copyright reserved.

Figures 9.2 and 9.4 reproduced with thanks to Rainhill Community Nursery, Merseyside, and copyright reserved.

Figure 9.5 reproduced with thanks to Anfield Infant and Early Years School, Liverpool, and copyright reserved.

Disclaimer: Some names (and identifying details) have been changed to protect the privacy of individuals/children.

Preface

'We've done this with Mister O'ara!' (said in a Yorkshire accent). I could not begin to tell you the number of times I heard this from children in my class. It didn't seem to matter what exciting activity I planned for the class, someone would utter, 'We've done this with Mister O'ara!' I was a young teacher in a junior school. Mark O'Hara was then an even younger teacher in our separate feeder infant school. Who was this Mr O'Hara and what did he do that was so inspirational?

As part of a research project in the 1980s, all teaching staff at our school were funded for four days to shadow a child for a day and then a member of staff for a day, in both the feeder infant school and the comprehensive school, where most of our children went. ICT was in its infancy. This was an enlightening experience and initiated plenty of discussion, among other things for me, about how we teach children science from the age of 5 to 16 and how to deal with transition. My visit to Brightside Infant School was my first real introduction to early years education. *And* I had the privilege of spending two days in Mr O'Hara's classroom! The environment was stunning. As I entered the classroom a large suspension bridge dominated the Victorian classroom, built by the children and Mr O'Hara. What I saw was inspirational. He certainly knew how to use the children's interest. He was able to make experiences memorable. He made links between science and everyday life. It was in this classroom that I first began to think about how very young children make sense of the world they live in. Mark O'Hara's teaching gave me my first great insight into what good early years practice could look like.

The first draft chapter I read for this book included a reference to Dr Mark O'Hara. We have both moved on from classrooms in the east end of Sheffield, but seeing his name took me straight back to those early years in our careers when we were learning our craft.

This book is intended for all those who would like to include more science in the early years curriculum. It is written by a variety of authors to inspire you to have a go. It provides a balance between the theory that underpins good practice and plenty of ideas of how you might put the theory into practice. There are many overlapping strands and themes, which are repeated in each chapter. For example we have included outdoor provision as a strand running through each chapter rather than a distinct chapter. We hope that this repetition will not be boring but will help you see how different voices explore the themes.

We have mainly drawn on current statutory documents for England to illustrate. However, this does not exclude those of you providing for children in Scotland, Wales, Northern Ireland or indeed other parts of the world. Draw on the ideas recorded here, which we believe are universal.

Thank you Mark O'Hara for helping me start the fascinating journey into science in the early years and for providing an inspiration to write this book.

Di Stead

1

Is science important in the early years?
Lois Kelly

Introduction

I wonder what you think or feel when you read the word 'science', particularly in relation to the curriculum in the early years foundation stage. Your views on whether science should feature within the early years will depend on how well your understanding of science 'fits' with what you already know about how young children learn. Do you view science as body of knowledge that has to be learnt as a set of facts and ideas or as a human endeavour to understand the world in which we live? Before reading any further take some time to reflect on whether science, through areas of learning such as Understanding the World, is relevant to young children's learning and development.

Task 1.1

Should we teach science to children in the early years?

1   Imagine you were asked to speak for this proposition in a debate. What reasons would you give?

2   Imagine you were asked to speak against this proposition in a debate. What reasons would you give?

Do the reasons you have given for teaching science to young children outweigh the reasons you have given against?

Your view and experience both of science and of how young children learn are likely to have influenced the reasons you have given. As you read the reasons given by associate teachers consider what has influenced their ideas. Compare the reasons given by the associate teachers with your own reasons; how were they similar and how were they different?

Reasons given for teaching science in the early years

- Young children are naturally inquisitive and science is all about finding out about the world; science encourages exploration and investigation.
- Science gives young children a better understanding of the world around them.
- Young children like to know how or why and science encourages them to question and test how things work.
- Children are introduced to science methods, techniques and concepts that they will learn about later on.
- Science helps children to understand their bodies, which underpins ideas about keeping healthy.
- Science is an interesting subject that engages young children.
- Children's vocabulary is extended and developed as they acquire scientific terms.
- Science has strong links to other areas of the curriculum.
- Science is a National Curriculum subject.

Reasons given for not teaching science in the early years

- Science is a complex subject and science concepts are too difficult for young children to fully understand.
- Young children cannot engage with the skills of science enquiry, measuring, recording and hypothesizing.
- Science knowledge is not needed to help early development.
- Young children do not have the vocabulary to express science concepts.
- Teachers may not have the necessary knowledge to teach science effectively.
- Science ideas may conflict with religious beliefs, which may confuse young children.
- It is more important to develop young children's literacy and numeracy skills.
- It takes too much time and effort to set up science activities.
- There are too many health and safety concerns.

What is science?

There is a consensus among science educators that science is a way of finding out about our world (Brunton and Thornton 2010: 1), involves providing explanations and searching for relationships between events based on sound evidence (Sharp et al. 2011: 7); and aims to learn more about and understand better objects, materials, living things and phenomena we experience (Wenham and Ovens 2010: 9). Characteristics of science highlighted by Harlen (2006: 34) include:

- that it is about developing an understanding of our world
- that it is a human endeavour which is dependent on human creativity and imagination to collect and interpret evidence
- that science theories change in the light of new evidence.

Beeley (2012) puts this into an early years context and suggests that for young children science is a way of making sense of the world through purposeful play. You only have to observe babies and young children for a relatively short time to notice that from a very early stage in their development they are continually exploring their world and developing scientific ideas (Johnston 2005: 1).

Early years curriculum frameworks in the UK recognize that science is one way in which children learn about the world they live in by including an area of learning that has the potential to develop children's science reasoning and thinking as well as learning about specific science ideas. In England this area of learning is Understanding the World (DfE 2014); in Northern Ireland it is The World Around Us (CCEA 2006); in Wales, Knowledge and Understanding of the World (DCELLS 2008); while in Scotland science is a specific curriculum area in the Curriculum for Excellence (Learning and Teaching Scotland 2009) for all levels from 3 to 18. While acknowledging that there is an argument for areas of learning in the early years curricula which encompass a fairly broad range of subject areas because young 'children's learning does not fit into neat subject categories' (CACE 1967, para 55) and that science is just part of children's learning about the world in which they live (Rogers 2012: 185), there is also concern that unless particular attention is paid in the early years to the scientific nature of children's explorations of their world (Esach 2011) a window of opportunity for expanding children's scientific thinking could be missed. Furthermore, Beauchamp (2013: 289) argues that each subject gives children a unique perspective for learning about the world, so learning how to work scientifically extends children's problem solving strategies. He suggests that early years practitioners need to consider when children should be a scientist or a geographer, for example.

A further argument for developing children's scientific thinking in the early years is that science also develops skills that children will need to use to function effectively as citizens in the twenty-first century. These skills include the ability to think critically, analyse information, apply knowledge to new contexts, understand new ideas, solve problems and communicate their ideas effectively (Yelland et al. 2008: 3). Those of us who have lived through the twentieth century are well aware of the advances that occurred in both science and technology throughout that time. My father, born before the end of the First World War, was an adult when Alexander Fleming discovered penicillin and had retired before personal computers or the Internet were features of everyday life. Similarly, we cannot imagine the developments in science that will occur during the lifetime of children born today. In Chapter 8, Eleanor Hoskins considers how technology opens doors into new worlds which children can explore.

Science in the early years

By the time children start more formal education in an early years setting they already have some principled and sophisticated science ideas. By the time children are 1 year old they know the difference between things that are living and things that are not living and experience has also taught them about cause and effect (Harlen 2010). This fact that was so obvious to one mum I was talking to recently that she wondered why it needed to be mentioned: 'Of course Felicity [aged 2] knows the difference between a doll and a baby.' You only have to watch babies and toddlers for a short while to realize that they soon develop an understanding of cause and effect. Before their first birthday children have learnt that if they drop something from a pram or their highchair it falls to the floor *every* time. They learn that a force is needed to make objects move, whether this is pushing or pulling a toy or being pushed on a swing. Furthermore, from their early experiences of making toast, eating an ice cream or chocolate and planting seeds they learn about change. In the following example, a reception class child uses what they know about solids and liquids to try to explain that a powder is not a liquid:

> It's something like a kind of liquid, but it isn't liquid, it's talcum powder. It goes fast like vinegar and it's not a solid because you can put your finger through it. It's a bit solid, but it isn't. You can break it, but it isn't water. You can put your finger through it, but it's not like a liquid. A liquid feels wet and this doesn't.
>
> (Russell et al. 1998: 120)

Task 1.2

Choose one of the following and make a list of the science ideas that children gain experience of from a very early age:

- **Bath time**
- **Meal time**
- **Playing in the garden**
- **The family pet.**

How young children construct meaning

This discussion is based on the understanding that young children actively seek to make sense of their world based on their own practical experiences and their interactions with others. The Tickell report (2011: 91) points out that young children actively think about the meaning of the activities they engage in and this leads to the creation of theories to explain the world in which they live (Worth 2000). A feature of the theories that children create is that they put together fragments of knowledge as they try to make sense of their world (de Bóo 1999: 85). As a result, they may then develop an explanation which makes sense to them but which may not be scientifically accurate; however, Worth (2000: 26) points out that these explanations are logical and based on both evidence and children's direct experience and provide an insight into their thinking. The following examples illustrate this:

> Final year BEd students were asked to find out how young children explained the change from day to night. The explanation given by one child in a reception class was:
>
> **'At night God puts on his special gloves and moves it [the sun] out of the way.'**

Unpicking this idea we can identify the different fragments of knowledge that led to this plausible explanation for what happened to the sun at night. This child knew he could not see the sun at night and I suspect that when he could not find a toy or a book it was because it had been 'moved out of the way' by a 'helpful' person tidying up. He also knew that the sun is hot and has observed adults 'putting on special gloves' when taking something out of the oven.

Figure 1.1 shows how another child has interpreted her observation of the apparent movement of the sun behind a hill to explain what happens to the sun at night-time (Figure 1.1). How has this child's personal experience informed her drawing and 'explanation'?

The idea that children (and adults) actively construct knowledge and understanding and have ownership over their learning is a central tenet of the constructivist view of learning and is supported by cognitive psychology. Throughout this discussion about how children construct meaning is the understanding that play and exploration are central to the way young children learn and are one of the characteristics of effective teaching and learning (Tickell 2011). De Bóo (1999: 7) points out that the skills children use when playing, exploring and investigating are fundamental to science and that the skills for working scientifically, which are integral to learning science concepts (Gelman and Brenneman 2004), are derived from them. A key feature of this approach is that scientific thinking, the process of finding out, is central to developing children's understanding of science concepts (Wilson 2007).

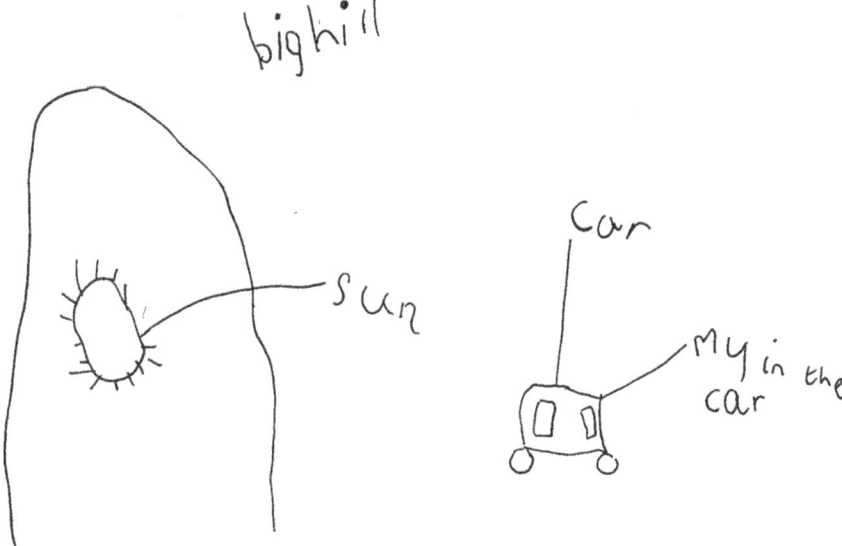

Figure 1.1 'The sun goes behind a big hill at night'.

The constructivist approach values children's existing knowledge and experiences uses these as a starting point for learning and then provides experiences that take children further along their learning pathway (Gelman and Brenneman 2004; Tickell 2012: 91). It is tempting to imagine this learning pathway as following a well-defined and very specific route, but this assumes that all children are interested in the same thing and that their learning follows identical routes, which contrasts with the widely accepted view that 'children develop and learn in different ways and at different rates' (DfE 2012). It may be more helpful to imagine learning as constructing a web where related activities are the individual strands in the web (Tickell 2012: 85) or like piecing together a jigsaw puzzle (Harlen and Qualter 2009). De Bóo (2006: 125) and Johnston (2005: 182) note that learning is more effective when children are given time to explore a chosen science concept using a range of related examples because this enables them to progress from applying their knowledge in familiar contexts to developing science ideas that are more widely applicable (Harlen and Qualter 2009). It is worth noting here that young children hold on tenaciously to the theories they develop because these are grounded in their experience and work in their daily lives (Worth 2000: 28). To help associate teachers think about how children learn and how children's ideas are resistant to change, Di Stead uses a jelly analogy. Imagine a jelly made in a round bowl then turned out onto a plate. If hot water is poured over the jelly it forms rivulets in the jelly. From a very early age children make sense of the world based on their experience; these rivulets represent their early ideas. Any more water poured over the jelly tends to run down the existing rivulets. In fact, it is quite difficult to get water not to run down the rivulets. Similarly, young children (and adults) tend to fit new evidence to their existing theories because it takes effort to change ideas that are grounded in our daily experiences. Another factor to take into account is that young children develop rules based on limited evidence which can be resistant to change (de Bóo 1999: 84; Evangelou et al. 2009: 5). Therefore providing a range of experiences supports children's learning by ensuring that they have a broader evidence base from which to develop these rules. Gelman and Brenneman (2004) point out that hurried exposure to too many disconnected science ideas fails to provide the opportunity for children to develop a rich understanding of science ideas. In Chapter 2, Kath Orlandi explores the need to give children time for sustained thinking. Furthermore, giving children both the time and the opportunity to make mistakes, and to repeat and revisit activities which they find engaging reinforces their learning (Brunton and Thornton 2010: 13; Worth 2000). In Chapter 5, Faith Fletcher discusses the role of continuous provision in providing opportunities for children to 'revisit, examine and retest' ideas.

Another reason for encouraging children to explore a science concept in a variety of different contexts is that it takes account of the children's differing interests by providing different 'entry points' for learning (Gelman and Brenneman 2004). Careful thought and planning are needed to provide 'entry points' which build on children's current experiences and interests. In Chapter 4, Linda Atherton explores some starting points and 'hooks' that generate interest and stimulate curiosity. Providing children with memorable experiences is of value not only for their science learning in the early years foundation stage but also because these experiences lay the foundations for the more complex science they learn later on. For example, observing the colours of the rainbow on the surface of a CD provides children with an early experience that white light is formed from a spectrum of different wavelengths of light, which can then be enhanced when they play with coloured spinners. Adding colour, using a spinner and noticing something 'brighter' gives a memorable experience that can be drawn on much later in school learning. Similarly, wearing glasses with coloured filters and noticing that yellow glasses seem to make all yellow things look brighter may be remembered and then can be explained later (Figure 1.2).

A discussion of the constructivist approach to learning also needs to consider the role of language in children's learning. As de Bóo (1999: 84) points out, children construct their understanding by a combination of practical activities and purposeful talk. One of the reasons that the associate teachers gave for not teaching science in the early years was that young children do not have the scientific vocabulary to express science concepts accurately. However, it is important to recognize that language is integral to children's learning and that

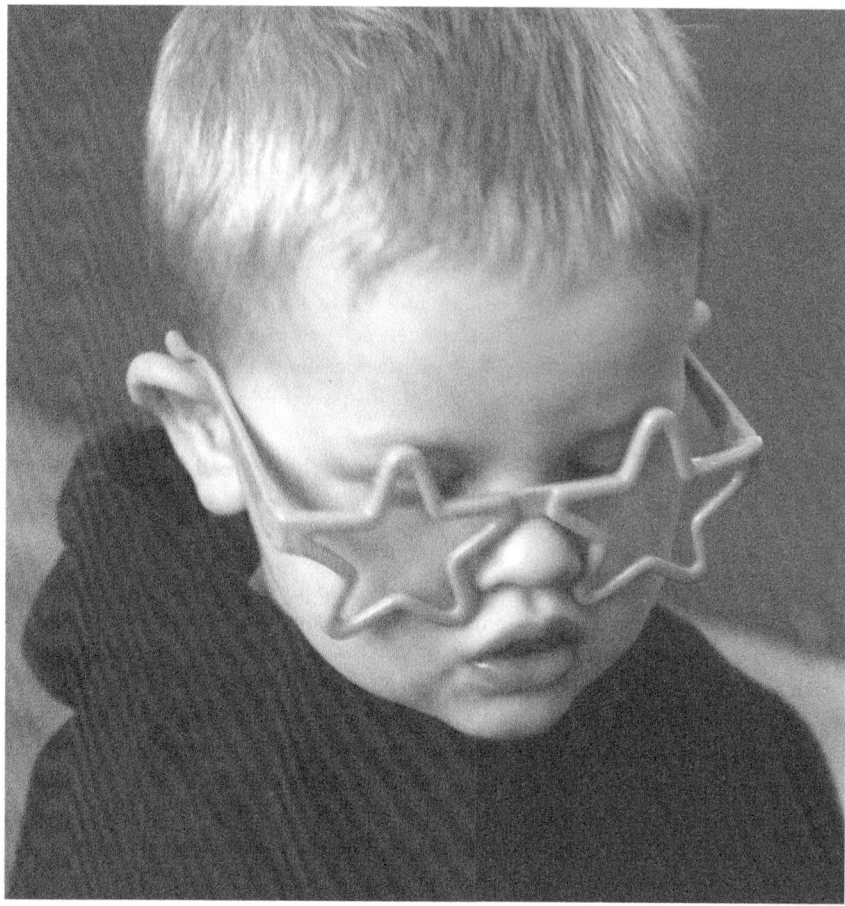

Figure 1.2 Wearing coloured glasses makes some colours seem brighter.

modelling the correct use of science language and vocabulary is simply another way of helping young children make sense of their world (Beeley 2012: 7; Brunton and Thornton 2010: 13). When stressing the importance of valuing children's natural ability to learn new words, Gelman and Brenneman (2004) talk about not 'cheating' on vocabulary that is relevant to specific science concepts and, as Harlen and Qualter (2009: 107) point out, adults need to introduce new science vocabulary at the appropriate time and ensure that it relates to the child's experience of a science event or phenomenon. Research into misconceptions in science has shown that misconceptions can arise because words take their meaning from the context in which they are used. It is quite common to hear both young children and adults in an early years setting talk about objects 'sticking' to a magnet rather than objects 'being attracted' to a magnet. This not only impoverishes children's language but it could lead to children developing a mental image of magnets containing a 'magic glue' (Nuffield Primary Science 1993: 64) rather than beginning to understand that magnetism is a force. Using the term 'see-through' to describe transparent can seem like a useful simplification; however, it reinforces the idea of the 'active' eye rather than the scientific explanation, which is that the eye is passive and we see because light enters our eyes. Developing a secure understanding of the science that children are gaining experience of and its related vocabulary gives adults the confidence to intervene sensitively to extend children's learning.

Beeley (2012) identifies three categories of scientific language that early years adults should model:

- The language of science processes or concepts – for example, *dissolving, melting, germination, growth, change*
- The language of science exploration and investigation (working scientifically) – for example, *questioning what if . . .? How . . .?; observing, predicting, compare and contrast, recording*
- The language of science equipment and objects.

Talk not only helps to develop children's understanding of science ideas, it also helps children develop a view of themselves as scientists and as a consequence their exploratory play becomes more scientific. (Howe and Davies 2010: 157–160). Babs Anderson discusses the role of talk in more depth in Chapter 3.

Key points

- **Science concepts, language and working scientifically are interconnected and mutually support children's learning (Gelman and Brenneman 2004; Harlen and Qualter 2009).**
- **Children need to be given a variety of experiences related to the chosen science concept and time to revisit these experiences.**

Nurturing children's curiosity

A key finding in the Ofsted science survey *Maintaining Curiosity* (Ofsted 2013: 16) was that early years foundation stage teachers nurtured children's curiosity to learn science by capitalizing on children's interests, providing them with the time and resources needed to explore and investigate and guiding them in developing basic skills. Implicit in this finding was that early years practitioners recognized that adult intervention was needed to support children's learning.

In this section, we argue that good understanding of science relevant to the early years coupled with a positive attitude to science enables adults to inspire young children, to develop positive attitudes to science and to nurture young children's natural curiosity to learn science, which may then lead them to study it later in life (Czerniak and Mentzer 2013; Spektor-Levy et al. 2013). While good science subject knowledge is important, understanding *how* children learn science has a significant impact on what children learn (Fleer 2009). Adults who are attuned to science are more likely to act as a guide and motivator (Johnston 2005: 167), recognizing science in spontaneous events and using these to develop children's understanding. They are also more likely to act as positive role models for the children, modelling the behaviours that encourage a scientific way of thinking and reasoning. Adults who had a 'scientific purpose' in mind supported children's learning using a variety of tools including talk, books and resources (Fleer 2009: 1084). Having a good understanding both of science concepts that are appropriate for the early years foundation stage and of how children's ideas develop over time helps adults to plan a range of relevant experiences which extend children's understanding of the science related to a particular theme or topic. In addition, adults who have a good understanding of science are more likely to use their subject expertise to develop children's understanding of science ideas by introducing the 'correct' science explanation as one of the ideas to be considered (Brunton and Thornton 2010: 13) and to know when or when not to launch into an explanation. In addition, they are more confident when modelling the correct use of science vocabulary and ensuring that the children in their setting have a balance of experiences across all areas of science. In Chapter 6, Jessica Baines Holmes illustrates how student teachers effectively used their science subject knowledge to support children's science learning in role-play areas. Similarly having good science subject knowledge enables adults to maximize children's science learning in continuous provision and will inform the choice of resources that supports this learning. (See Chapters 5 and 7.) Developing both the science subject knowledge and the pedagogical knowledge that are appropriate for the early years foundation stage is an ongoing process and is a characteristic of good early years practitioners who take responsibility for their own professional development. Alexander (2003) argues that expertise develops as practitioners become accustomed to particular ways of working and the problems associated with an area of learning.

Wenham and Ovens (2010: 11) point out that 'knowing how to' work scientifically is an essential part of learning to be scientific and also supports the development of children's knowledge and understanding of science ideas. Developing expertise in this aspect of science enables adults to work alongside children as a co-investigator and a role model (Campbell 2012; de Bóo 1999: 41–45), developing the children's ability to work and think scientifically. Working as a co-investigator provides opportunities for the adult to nurture children's natural curiosity by both valuing the questions children ask and asking questions that guide children's explorations and maximize their learning (de Bóo 1999: 32) and at the same time help them experience 'the pleasure of finding things out' (Loxley et al. 2010: 3–11). In Chapter 3, Babs Anderson explores ways that the questions adults ask help children to explore their science thinking and points out that the quality of these questions is influenced by the level of the teacher's science subject knowledge and that they are more effective when they become a feature of planning (Goldsworthy 2011: 75). Faith Fletcher in Chapter 5 provides some helpful suggestions for the types of question that encourage children to work scientifically. It may seem counterintuitive to plan for questions that initiate children's explorations and investigations in an environment which values children's interests, but good preparation will enable adults to respond sensitively and appropriately to children's questions. However, with practice and by reflecting on what works and what does not work adults can develop their questioning expertise.

Responding appropriately to children's questions does not necessarily mean telling children the answer every time they ask a question and it is unrealistic to expect to know the answers to all the questions children may ask. Beeley (2012) reassures early years practitioners that personal science knowledge is constantly evolving and involves building on existing knowledge. When adults are faced with questions to which they do not know the answer she recommends that they should use them as starting points for investigations. Harlen and Qualter (2009: 150) point out that, to nurture children's natural curiosity to learn, adults need to value the question they ask by considering what type of response is most appropriate. A 3-year-old who asks 'Why is there no water in the river?' and 'Where does the water go?' at low tide does not want a detailed explanation of what causes the tides, but is showing that they have noticed that at different times of the day the river is either full of water or empty. Harlen and Qualter point out that even if the adult does know the answer to this type of question, the child would probably not understand the answer and by simply acknowledging the child's interest the adult nurtures a sense of wonder. When working alongside children as a co-investigator, adults are well placed to encourage children's questions by wondering about the world (de Bóo 1999: 19) and modelling the behaviours which help the children find the answer. An example of this is the adult in Rainhill Community Nursery who wondered with the children why the birds were tweeting (see Chapter 9).

Task 1.3

Harlen and Qualter (2009: 150) identify five types of question that children might ask:

1   Comments expressed as questions
2   Philosophical questions that science enquiry cannot provide an answer for
3   Requests for simple facts, for example a name or a definition
4   Requests for an explanation that would be too complex for a child to understand
5   Requests for an explanation that a child could find through science enquiry.

Sort these children's questions using the categories above:

* Are all seeds the same?
* Why has this piece of apple turned brown?
* What is this flower called?
* Why do balls bounce?
* Where does a woodlouse live?
* Why am I taller than Georgia?
* Why does a puddle get smaller?
* How long does it take ice to melt?
* Why is the sky blue?

To conclude, adults who have a positive attitude to science and share their enthusiasm and curiosity with children will inspire positive attitudes to science in the children in their setting. If you believe that science in the early years sows the seeds from which children grow their knowledge and understanding of science concepts then you will recognize the importance of ensuring that the science experiences you plan for them are informed by a good understanding of science, working scientifically and the correct use of scientific language that supports this learning. Two books which explain the background science knowledge relevant to the early years are *Science in the Early Years: Building Firm Foundations from Birth to Five,* by Pat Brunton and Linda Thornton (2010) and *Talking and Doing Science in the Early Years: A Practical Guide for Ages 2–7,* by Sue Dale Tunnicliffe (2013).

Key points

Developing good background science subject knowledge, relevant to the early years foundation stage, is important because:

- it informs planning for a range of relevant activities that develops children's knowledge and understanding of any particular science topic
- adult intervention and the quality of the questions are influenced by their level of science knowledge and
- children's natural curiosity to learn science is more likely to be nurtured when adults have a good level of science background knowledge.

Exploring good practice

The contributors to this book explore young children's natural curiosity to learn about their world and the importance of helping children develop their science thinking. The four guiding principles of the *Statutory Framework for the Early Years Foundation Stage* (DfE 2014) in England that every child is unique, that children will develop well in enabling environments, that positive relationships help them to become strong and independent and that children learn at different rates are themes common to each chapter. These principles also inform the Foundation Stage in Northern Ireland (CCEA 2006) and Scotland (Scottish Executive 2007) and the Foundation Phase in Wales (DCELLS 2007).

When exploring the development of an enabling environment that inspires science in the early years foundation stage, the contributors discuss opportunities for learning both inside and outside the classroom. Children's experiences of science concepts inside and outside the classroom are both qualitatively and quantitatively different. I have been in many early years settings where a 'cave' has been set up in the role-play area so that children can experience the fact that we need light to see. Taking the children outside at different times of year to explore the difference between sunny and shady places extends their learning about light as they not only explore differences in the intensity of light but also feel the difference in temperature. A more adventurous setting developed a cave in their grounds to enhance children's science (Rowe and Humphries 2004: 21).

Contributors also explore the role of the adult to inspire science in the early years by introducing children to the language that helps them to explain their ideas, developing role-play areas to explore a science topic, enhancing continuous provision, providing interesting starting points for their learning and making available the resources that enhance their learning and the opportunities that technology provides to open doors to new worlds.

References

Alexander, P. (2003) The Development of Expertise: The Journey from Acclimation to Proficiency. *Educational Researcher* 32(8): 10–14.

Beauchamp, G. (2013) Understanding the World, in I. Palaiologou (ed.) *The Early Years Foundation Stage: Theory and Practice*. London: Sage.

Beeley, K. (2012) *Science in the Early Years: Understanding the World through Play-Based Learning*. London: Featherstone Education.

Brunton, P. and L. Thornton (2010) *Science in the Early Years: Building Firm Foundations from Birth to Five*. London: Sage.

CACE (Central Advisory Council for Education) (1967) *Children and Their Primary Schools* (The Plowden report). London: HMSO.

Campbell, C. (2012) Identifying some Issues in Professional Learning in Early Childhood Science. *Journal of Emergent Science* 3: 15–21.

CCEA (2006) *Northern Ireland Curriculum: Understanding the Foundation Stage* (retrieved from www.nicurriculum.org.uk/docs/foundation_stage/UF_web.pdf).

Czerniak, C.M. and G. Mentzer (2013) Early Childhood Science: A Call for Action. *School Science and Mathematics* 113: 157–158.

Dale Tunnicliffe, S. (2013) *Talking and Doing Science in the Early Years: A Practical Guide for Ages 2–7*. Abingdon: Routledge.

DCELLS (Department for Children, Education, and Lifelong Learning Skills) (2008) *Framework for Children's Learning 3–7 in Wales* (retrieved from http://wales.gov.uk/dcells/publications/policy_strategy_and_planning/early-wales/whatisfoundation/foundationphase/2274085/frameworkforchildrene.pdf?lang=en).

de Bóo, M. (1999) *Enquiring Children, Challenging Children*. Buckingham: Open University Press.

de Bóo, M. (2006) Science in the Early Years, in W. Harlen (ed.) *ASE Guide to Primary Science* (3rd ed.). Hatfield: ASE.

DfE (Department for Education) (2014) *Statutory Framework for the Early Years Foundation Stage* (retrieved from www.gov.uk/government/uploads/system/uploads/attachment_data/file/299391/DFE-00337-2014.pdf).

Esach, H. (2011) Science for Young Children: A New Frontier for Science Education. *Journal of Science Education and Technology* 20: 435–443.

Evangelou, M., K. Sylva, M. Kyriacou, M. Wild and G. Glenny (2009) *PEEP Research Birth to School: Research Report DCSF –RR176* (retrieved from www.gov.uk/government/uploads/system/uploads/attachment_data/file/222003/DCSF-RR176.pdf).

Fleer, M. (2009) Supporting Scientific Conceptual Consciousness or Learning in 'a Roundabout Way' in Play-Based Contexts. *International Journal of Science Education* 31(8): 1069–1089.

Gelman, R. and K. Brenneman (2004) Science Learning Pathways for Young Children. *Early Childhood Research Quarterly* 19: 150–158.

Goldsworthy, A. (2011) Effective Questions, in W. Harlen (ed.) *ASE Guide to Primary Science* (4th ed.). Hatfield: ASE.

Harlen, W. (2006) *Teaching, Learning and Assessing Science 5–12*. London: Sage.

Harlen, W. (ed.) (2010) *Principles and Big Ideas of Science Education*. Hatfield: ASE.

Harlen, W. and A. Qualter (2009) *The Teaching of Science in the Primary School* (5th ed.). Abingdon: Routledge.

Howe, A. and D. Davies (2010) Science and Play, in J. Moyles (ed.) *The Excellence of Play* (3rd ed.). Maidenhead: Open University Press.

Johnston, J. (2005) *Early Explorations in Science* (2nd ed.). Maidenhead: Open University Press.

Loxley, P., L. Dawes, L. Nicholls and B. Dore (2010) *Teaching Primary Science: Promoting Enjoyment and Developing Understanding*. Harlow: Pearson.

Nuffield Primary Science (1993) *Key Stage 2 Electricity & Magnetism Teacher's Guide*. London: Collins Educational.

Ofsted (2013) *Maintaining Curiosity: A Survey into Science Education in Schools* (retrieved from www.ofsted.gov.uk/resources/130135).

Rogers, R. (2012) *Planning an Appropriate Curriculum in the Early Years: A Guide for Early Years Practitioners and Leaders, Students and Parents*. London: David Fulton.

Rowe, S. and S. Humphries (2004) The Outdoor Classroom, in M. Braund and M. Reiss (eds) *Learning Science Outside the Classroom*. London: Routledge Falmer.

Russell, T., K. Longden and L. McGuigan (1998) *SPACE Project Research Report: Materials*. Liverpool: Liverpool University Press (retrieved from www.nationalstemcentre.org.uk/elibrary/resource/4541/space-project-research-report-materials).

Scottish Executive (2007) *A Curriculum for Excellence: Active Learning in the Early Years* (retrieved from www.educationscotland.gov.uk/Images/Building_the_Curriculum_2_tcm4-408069.pdf).

Sharp, J., G. Peacock, R. Johnsey, S. Simon, R. Smith, A. Cross and D. Harris (2011) *Achieving QTS: Primary Science Teaching Theory and Practice* (5th ed.). Exeter: Learning Matters.

Spektor-Levy, O., Y.K. Baruch and Z. Mevarch (2013) Science and Scientific Curiosity in Pre-School: The Teacher's Point of View. *International Journal of Science Education* 35(13): 2226–2253.

Tickell, C. (2011) *The Early Years: Foundations for Life, Health and Learning* (retrieved from www.gov.uk/government/publications/the-early-years-foundations-for-life-health-and-learning-an-independent-report-on-the-early-years-foundation-stage-to-her-majestys-government).

Wenham, M. and P. Ovens (2010) *Understanding Primary Science* (3rd ed.). London: Sage.

Wilson, R. (2007) Promoting the Development of Scientific Thinking, *Early Childhood News* (retrieved from www.earlychildhoodnews.com/earlychildhood/article_view.aspx?ArticleId=409).

Worth, K. (2000) The Power of Children's Thinking, in *Foundations: A Monograph for Professionals in Science, Mathematics and Technology Education. Volume 2: Inquiry* National Science Foundation (retrieved from http://opas.ous.edu/Committees/Resources/Publications/Foundations_Inquiry_nsf99148.pdf#page=29).

Yelland, N., L. Libby, M. O'Rourkes and C. Harrison (2008) *Rethinking Learning in Early Childhood Education*. Maidenhead: McGraw-Hill.

2
Developing budding scientists
Kathleen Orlandi

This chapter begins by looking at the nature of children's learning in relation to science and how they have fundamental development needs to achieve a sense of belonging to the world into which they are born. It continues by suggesting ways in which these needs can be met through an enabling environment that encourages curiosity and ingenuity.

The empiricist view of childhood is a deficit one in which the teachers need to fill the 'gaps' in knowledge and skills. An interactionist viewpoint would be that the child develops through interactions between his own mental structures, the environment and the thoughts and ideas of other people; this would place the child as a budding scientist, exploring his world and trying to make sense of it. The social-constructivist view would be that a more knowledgeable other would be encountered within this environment and support the child in developing scientific concepts and skills.

Science should not be confined to the secondary school classroom or primary classroom. The essential experiences that help develop budding scientists should be there to establish a positive attitude to risk taking, enquiry, investigation, trial and error, testing and discovery from an early age. Children are naturally curious. Quite frequently academics use the example of snowfall and the teacher who admonishes the children for staring out of the window rather than paying attention to his or her words as an example of adults' lack of understanding of children's ability to be in awe of a phenomenon; or indeed of not knowing how to follow up on that curiosity and enthusiasm. Most parents will be familiar with the experience of having to draw a reluctant child away from something in which they are engrossed, in order to deal with their everyday tasks such as shopping. From birth babies use all their senses to explore their environment; they touch, grasp, look, taste and listen to the environment. It is their instinctive curiosity that motivates them to become mobile in order to explore and examine the immediate environment. It is natural that they want to be 'on the move' and to touch and explore yet we often prevent children from doing this because we fear for their safety. I have enormous sympathy for young children being taken on a supermarket trip, where they are not allowed to touch anything and where, often, parents can be heard being aggressive when the child cries or demands their attention. Young children were not made to be sedentary, otherwise they would not be able to explore and investigate, yet institutionally we expect them to remain still for unreasonable lengths of time. They have a basic need to be mobile, yet some teachers expect them to sit still on the carpet for periods of time that are unreasonable and ineffective.

In school, a play-based approach to learning, advocated by the early years foundation stage guidance, should provide a 'safe risk-taking environment that allows budding scientists to develop a scientific brain by exploring, investigating and inventing' (Orlandi 2012: 101). However, there are some key requirements if the environment is to enable the emerging scientific skills to be nurtured:

- Encouragement to explore and investigate
- Uninterrupted thinking time
- Access to the world beyond the classroom
- Provision for untidiness.

Encouragement to explore and investigate

Haywood (2004) warns that we should not rely on a child's natural motivation to explore and gain knowledge from birth, as this tendency is shaped by the people around the child, including the teachers. I do worry about how adults interpret guidelines. It is important that adults working with young children understand their role. I have visited schools where teachers believe that freedom in the flow of play requires little input from them, other than to provide the resources. Some think it is their role to step back and stay back. It is a complex and sensitive role that is required of teachers to support children in playful learning without interfering. Children need help in order to become autonomous in their learning. Claxton (2002) identifies four characteristics to foster in order to build the learning power of children: resilience, resourcefulness, reflectiveness and reciprocity. Too much adult interference or unrealistic expectations could result in a fear of failure on the part of the child. This would result in a reluctance to experiment and investigate.

Clearly the teachers' own knowledge, understanding and attitudes towards science are important. If teachers believe that science is simply a body of knowledge acquired and tested by others, they will not appreciate the significance and value of the children's explorations and investigations. However, if they 'regard scientific knowledge as shifting and tentative – inherently rooted in the "here and now" of everyday thing and events – early years science will appear as a natural component of young children's play' (Howe and Davies 2005: 157). During exploration children should be able to have a wide range of experiences so that they can make many connections and develop the complex neural networks associated with learning. Howe and Davies (2005) remark on children's more imaginative approaches to observation that may be holistic, rather than the focused attention teachers may draw to particular features. If teachers have a heightened awareness of the scientific nature of children's play they will see it as exploration and investigation of all kinds of phenomenon such as forces, gravity, plasticity, density, mass, light, colour, movement, energy, changing properties, evaporation, living things and many more.

Exploration and investigation should be of the whole environment, including other children and adults, so that the experiences are authentic, contextual and are 'socio-culturally relevant' (Robbins 2005: 6). Authenticity is vital so that children can see meaning and purpose in what they are doing. It is sometimes difficult to provide authentic life experiences within the confines of a school environment. However, it is possible, and indeed essential, to introduce elements of the 'outside world' if children are to be able to improve their knowledge and understanding of the world into which they are born. For example, old clocks, old telephones, weighing scales and similar devices provide opportunities for children to unravel the physics of such items. I was privileged to witness a child from a reception class dismantling an old clock for a long period and later on the same day recreating the clockwork parts and movements in the workshop area. It had gone unnoticed by the busy teacher but was a valuable learning experience. If there is construction work or renovation work happening in or near the school, the first concern will inevitably be health and safety, and also whether or not the activity will disrupt the children's education. Perhaps, within the parameters of health and safety, teachers of young children should see this as an opportunity for the children to observe and ask questions. There are many aspects of science that could be learned through this process including forces, energy, velocity, gravity, gradients, measures and materials. Incorporating apparatus such as spirit levels and string, and buckets with string to make pulleys, into the construction areas of the classroom will encourage the children to investigate the ideas they have encountered.

School visits are traditionally annual events and it is easy to understand the infrequency because of the organization, costs and risk assessments involved. Although a visit to the fire station is really exciting for the children, it is possible for the fire brigade to bring an engine to the school so that they can learn about all the equipment and how it operates. An example that demonstrates the knowledge gained from such a visit was when an Ofsted inspector queried a lesson plan involving a story of a 'hoist'. The inspector judged that this was too difficult a concept for young children until he learned that they had watched a hoist being used the previous week to load something onto a fire engine. It is also worthwhile finding out about the occupations of the parents and grandparents of the children. It is possible that there may be a nurse who could demonstrate equipment such as a stethoscope or thermometer in the context of his or her role. I discovered that the family of one of the children in my class went deep sea diving and invited his father in to demonstrate the equipment and answer questions. Questions are important and teachers can model the kind of questions children should ask when investigating, for example:

- What will happen next?
- How did that happen? Can we make that happen again?
- How could we make this faster/slower/larger/smaller/longer?
- Why do you think it did that?

Cooking and baking were once common activities in the classroom but sadly are not part of many children's experiences today. These activities provide opportunities to explore the properties of various ingredients, and to investigate how the properties change when mixed and when heated. Children can experiment with ingredients to invent their own recipes and evaluate these according to taste and appearance. Better still is to use ingredients that have been grown and nurtured by the children so that they begin to make connections between the climate, the earth, plants and our food. Conversations with the children during these activities support their developing grasp of the concepts as well as associated vocabulary. Not only does this provide authenticity it enables children to make connections. Too often activities provided by teachers are isolated and do not enable children to see the context or the connection with the world outside of school. Adams et al. (2004) found that activities planned to promote learning were sometimes impoverished and recommended authentic, stimulating and purposeful experiences from the real world to challenge and satisfy children. If children are free to explore, investigate, experiment and invent in their own time, they will be able to make connections between these experiences and previous ones, therefore gaining greater insight.

Key points

- **Children need encouragement to explore and investigate.**
- **Children need authentic and meaningful experiences.**
- **Children need to be free from the fear of failure so they will experiment.**
- **Children need encouragement to ask questions.**
- **Children need to be supported by teachers who understand the scientific nature of their play.**

Uninterrupted thinking time

In school, the daytime tends to be divided into recognizable slots. These usually fit in with a common start to the day around 9am, a mid-morning playtime, a lunch hour at midday, a mid-afternoon playtime and ending the day at approximately 3.30pm. There have to be set times within an organization such as a school. However, these episodes of time are often further divided into story time, literacy, numeracy, PE, school assembly and so on. The teacher plans how the time will be used, leaving the child having to fit in around the timetable devised by adults. May et al. (2006) refer to this as chunking of time. This often results in the children being deprived of sufficient time to explore, investigate and think. Pondering is an essential aspect of scientific enquiry, the interruption of which can be frustrating. In my own research, I found that the *only* aspect of school that the children verbally and visibly expressed frustration about was the abrupt halts to their investigations and explorations. Adams et al. (2004) also found that children were often called away from deeply involved play to attend adult-led activities that were less challenging. I suggest that there are three reasons why this happens. First, there is the perceived need for teachers to chunk the time to suit planned activities. Second, there is is the fact that they have to take breaks at the same time as the rest of the school to fulfil their part in the playground duty rota. Third, they do not fully appreciate the value of the learning that is taking place when the children are 'just playing'. The deeper in thought (deeper learning) they are, the more frustrated the children become when interrupted.

There are several reasons why children need to have long periods of uninterrupted time to think. They may become deeply engrossed in their activity and need time to ponder the new concepts; often this requires repetition and practice. They need to have opportunities to move between different parts of the environment and to ask questions in order to make connections with previous experiences. It requires time for the children to become engrossed and to 'think things through'. As adults we should not find this difficult to comprehend. Scientists need to pursue and persevere in their quest to be inventive and investigative, yet abrupt interruptions remove opportunities for the children to develop these traits. Could children's scientific potential go undiscovered because of inadequate opportunities?

'Trial and error' is associated with the scientist at work. Adult scientists return to an investigation many times, whether inventing or discovering. It is not unusual to walk away from an experiment then, having had time to think, often overnight, to come up with an idea to try or a solution when we return to the task. It is the same for children. Children need sustained periods of time to pursue their enquiries and to be able to return to them at some other point in time. Sadly, their work is often cleared away or the resources made available for other children. No wonder their resilience and perseverance is affected.

Sustained periods of uninterrupted time would allow children to verbalize their thoughts to others. This space for talk should not be undervalued. Articulating our thoughts to others helps us to clarify a developing concept for ourselves as well as the recipient. If children can share their thoughts with other children and adults in a reciprocal way the shared understanding will improve the progress of exploration, investigation and inventiveness. 'Sustained shared thinking' (Siraj-Blatchford et al. 2003), in which children share their thoughts over a period of time with all parties learning together, is essential for growth in knowledge and understanding. Language is the vehicle in which knowledge is shared and developed. A healthy classroom would be one in which children were engrossed in investigation of their environment and in which there were meaningful and purposeful conversations, sometimes involving adults, that increased the vocabulary, and therefore ability, to construct meaning within the parameters of their explorations and investigations. The development of language that supports scientific enquiry is best learned *through* purposeful scientific enquiry, as this would lead to 'richer language than might occur in artificially created situations' (Ellis 2005: 12, 13).

If sufficient time and, more importantly, sufficiently long episodes of time are available to the children it will give them time to observe and think. The term 'daydreaming' has often been used in a negative way, to suggest that a child is not concentrating. I suggest that it is a desirable intellectual activity in which children ponder and make connections between various experiences. We should allow for this sometimes rather than draw the child back to the adult agenda. This has echoes of the 'absent minded professor', who I suggest is not so much absent minded but has a busy brain that is wandering elsewhere making connections. I am sure that most teachers can recall children who show these traits.

Teachers also need time to observe the children. Unfortunately, some teachers feel compelled to teach focus groups when the other children are playing; this is understandable because of the targets they are expected to achieve. However, rich knowledge about the children can be gained by observing them over lengthy periods. Having had the luxury of observing children in a project without having teaching responsibility I noticed many aspects of their scientific enquiries that went unnoticed by the teacher. For example, one day a small child in a nursery class spent over an hour in solitary play examining the properties of sand. She looked at the grains. She closely examined their movement through a funnel and wheel (Figure 2.1). She repeatedly experimented with gently pouring sand onto one mound and was very persistent in her pursuit of a solution to the problem of the mound not gaining height, on account of the movement of the grains. She seemed to be oblivious to the noise and bustle around her and seemed not to hear the teacher calling an end to the morning session. She returned to the same task the following day and showed resilience and determination to complete his investigation. If she had not been given sufficient time to play this trait would not have been observed. Another example of a child persistently returning to an investigation is Callum in Chapter 5.

Figure 2.1 Closely examining the movement of dry sand through a wheel.

Achieving a balance between teacher-led and child-initiated activities is difficult. I suggest that small periods of time for children to choose and follow their own interests through play are of little value, as they do not enable the children to become engrossed in their activity. I recommend that fewer but longer periods of time for child-initiated activity, in which teachers observed and sometimes provided sensitive support and involvement, would lead to an environment suitable for the development of budding scientists.

Key points

- Children need time to think and ponder.
- Children need time to articulate their thinking during their investigations and explorations.
- Uninterrupted time periods of adequate duration are needed for children to develop the skills of scientific enquiry.
- Uninterrupted periods of time enable better development of knowledge and understanding of the world.
- Fewer but longer episodes of child-initiated activity or play are better than several brief impoverished episodes.

Access to the world beyond the classroom

As mentioned earlier in the chapter, children need to access the world into which they are born, of which the classroom is a very small part; they spend a disproportionate amount of time in this part of their world. Therefore if they are to explore and investigate this wider world they need access to it. It does not seem right that the only glimpse of this world is through adult constructs of it through books and formal teaching. The difficulties of taking children into this wider world have already been alluded to but we often do not make the best use of what is accessible. All schools have an outdoor environment which, even without enhancement, can provide resources for learning about the world (Figure 2.2). When questioned about developing the outdoor environment for learning, space and costs are often cited as reasons for underdevelopment. However, some of the best examples I have seen have been in confined spaces and without economic investment. The elements of the weather are freely available and provide opportunities for scientific discussions. For instance, the children can discuss how the temperature alters when the sun appears or disappears behind clouds. They can conduct an experiment to discover which weather conditions are best for drying washing and discuss the reasons why a warm still day might be less effective than a cool windy day, thereby exploring the concepts relating to evaporation. Similarly, in addition to the enjoyment of making snowmen and snowballs, the children can discuss the properties of snow and ice. For example, they might investigate how the snow forms into shapes and how the properties change when under pressure or through a change of temperature. Simple tools to manipulate the snow and hand lenses will encourage investigation.

In some schools, the outdoors provides more space than the indoors for children to wander, spend a little time on their own exploring, imagining and pondering (Harding 2005). For children to develop scientific skills of enquiry they need to be encouraged to take risks, both in terms of practical activity and in terms of fulfilling

Figure 2.2 The outdoors is an ideal environment for children to learn about factors that support life.

the perceived requirements of the teacher. The outdoor environment provides greater opportunities for this to happen, partly because of the space and also because the children will feel less under the watchful eye of the teacher. Children who worry about negative judgements of others will play safe, and even confident children will

Figure 2.3 Minibeasts and other wildlife provide resources for children to explore.

rein in their adventurous spirit (Claxton 2002). Waters and Begley (2007) found that, more than just allowing it, the adults provide for risk taking outdoors.

The outdoors is an ideal environment for children to learn about the cyclical nature of life and factors that support life. Minibeasts and other wildlife provide resources for children to explore (Figure 2.3). It is better to study them in authentic situations rather than within the classroom in a laboratory fashion. Simple containers and hand lenses will add to the opportunities to investigate (Figure 2.4).

Children can grow plants for decoration and for food, even within small spaces. I witnessed a classroom with a small balcony which was filled with trailing strawberry plants, potatoes, runner beans and other vegetables. The abundance and variety in a small space was remarkable. By sowing and nurturing plants children learn about how life is supported and about the cycles and seasons involved. An excellent example of making the best use of the outdoors for growing plants was of a colleague who worked in an inner-city nursery. The children came to nursery dressed to be outdoors. They grew their own plants and harvested them. The vegetables were used in recipes which the children selected. They were encouraged to explore the flavours and textures and invent their own recipes using their own grown ingredients. Each stage of the cycle of growth was recorded from sowing seeds to cooking and compiled into books so that the children could

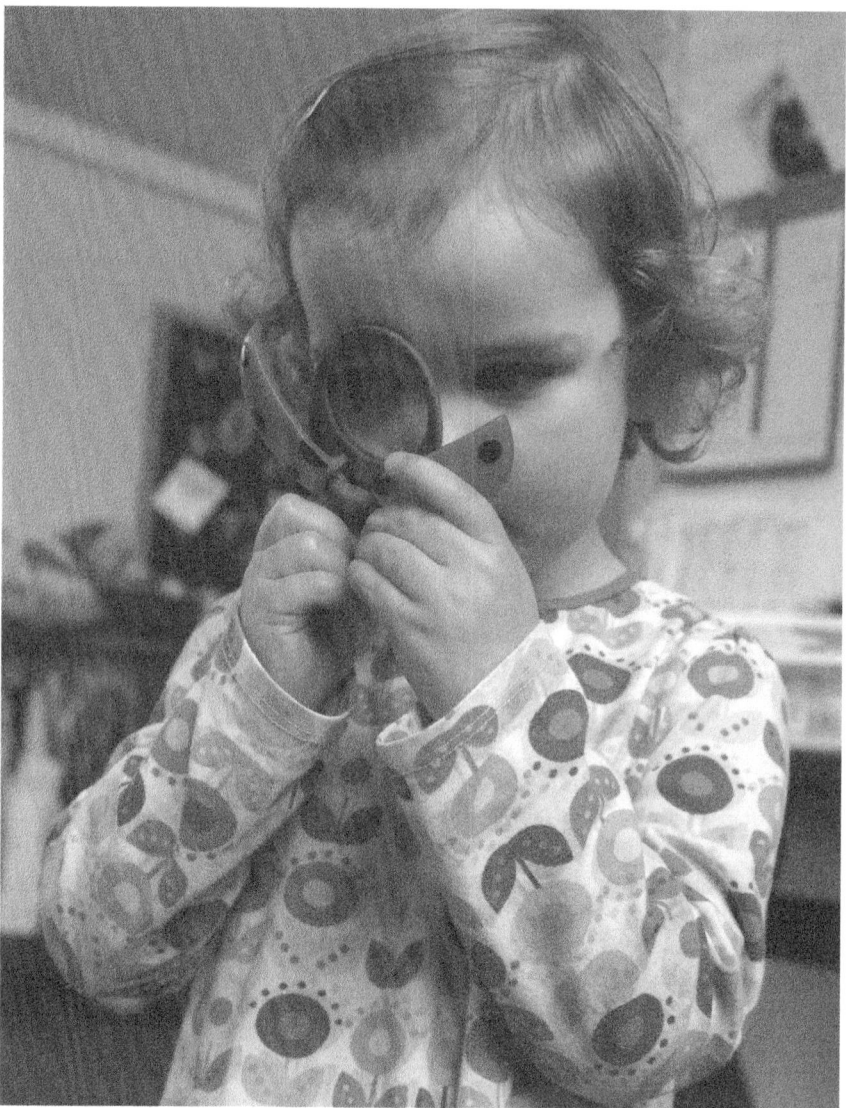

Figure 2.4 Hand lenses encourage observations and explorations.

recall the processes and sequence of events and make connections with previous experiences. It also encouraged them to articulate their developing grasp of some of the scientific concepts involved.

The outdoor space enables children to use large equipment, to explore mechanisms and investigate concepts such as velocity and gradient. Boxes, tubes and pipes are the basic resources needed. It will encourage the children to be inventive and resourceful in construction. The addition of wheeled toys, sand and water will encourage them to experiment with movement and processes. This environment in which children have space and can be noisy is more productive than the indoors in terms of motivation and learning as it is less restrictive. Children are more likely to take the necessary risks if the results of their activity are not going to be seen as disruptive. If children are able to take photographs or film footage of their activities, they will be able to bring the outdoors in for sharing and to reflect on and evaluate their work. As in the example in the previous paragraph, this encourages development of the language that will support the grasp of scientific concepts and further enquiry.

It is worth considering the rights of the child in the context of opportunities for them to develop their potential skills in science and particularly their rights to access the wider world, including the outdoors, in fulfilment of a basic development need to explore and investigate. The United Nations Convention on the Rights of the Child, ratified by the UK government in 1991, is about children's participation in decision making. They have a right to express their views and for their views to be given due weight. Children like being outdoors. Many children's need to access the outdoors is such that too much time in the classroom results in unsatisfactory experiences for them. We need to safeguard against children being prevented from having quality access to the outdoor environment and the opportunities it offers because of decisions made by adults.

Key points

- **The outdoors provides a good environment for budding scientists because they can take more risks.**
- **Teachers allow and encourage risk taking outdoors.**
- **Even without enhancement, the outdoors provides opportunities for budding scientists to explore the environment.**
- **The outdoors provides authenticity to the exploration of life processes.**
- **Outdoor space allows room for large equipment for children to explore and examine physical processes.**
- **Children have a right to access the outdoors.**

Provision for untidiness

Initial teacher education programmes often include a lecture on classroom management, time management and organization. Visitors to classrooms, including some school managers, would be impressed by a tidy, apparently well-ordered classroom. Indeed, good organization is a key skill for teachers, and a well-organized environment with everything in its place reflects that the teacher in charge takes some pride in this and cares. I can identify with this; some forms of untidiness result from a lack of care and dedication. However, this part of the chapter argues that we should not only allow some untidiness, but should provide for it.

When children's brains are making connections with previous experiences, this may happen in a way that does not appear logical to us. To illustrate this point I will share an example of something that happened in a reception classroom, which will be familiar to experienced teachers. A teacher was telling a story that involved a blue car which had broken down and whose driver had to walk a long way to get help. The teacher continued with this story, which involved various encounters with people. A child interrupted the storytelling, announcing that his dad had a new coat. The teacher nodded and went on with the story. A subsequent discussion with the child revealed that what had appeared to be a comment made out of context had indeed been connected to the story. His mind had wandered from the blue broken-down car, to a trip to Manchester when the train had broken down. The family had eventually arrived in the city and his dad had purchased a new coat. It is not unusual for children to join in conversations with apparently unrelated stories which, if time permits, will be found to be connected to the main focus.

When a child is playing – that is, exploring, investigating and inventing – they will be making connections to many kinds of previous experience. Within the classroom this may result in the child moving from one recognized area to another, and sometimes taking with them an object which is the subject of their thinking, in order to connect with a prior experience in the next location. They may be exploring a concept about the properties or functions of the object. This type of play, in which the child is moving freely between various parts of the environment and various experiences, was labelled 'free-flow play' by Bruce (1991). Free-flow play involves being imaginative, creative, original and innovative and includes first-hand experiences of exploration and discovery. It therefore lends itself to the development of scientific skills and knowledge. To those unfamiliar with the notion of play as a form of scientific enquiry, the classroom scene may appear to be untidy, but then the brain does not make connections in a tidy fashion. For those of us who have had the good fortune to be able to observe children at play, without the added responsibility of supervising the class, it becomes obvious that high-level learning is taking place. Gentle questioning when the opportunities arise give interesting insights into the lines of enquiry pursued by the young scientists. Some researchers find that uncertainty and disorder can lead to richer understanding and that the brain is designed for chaos. Jensen (1995: 85) suggests that we should encourage an 'orchestrated chaos' to allow more variety and reality into learning contexts.

Daydreaming is an essential occurrence if children are to ponder on the concepts they are trying to grasp. The 'relaxation of constraints' is an attribute of daydreaming identified by Mueller and Dyer (1985) allowing the exploration of possibilities which may not have otherwise been pursued. They assert that some of these possibilities may lead to new solutions to problems. Previous experiences may help to reach an understanding or may inform the next steps of the enquiry. Daydreaming is seldom tolerated during adult-led activities and is seen as 'not paying attention' to something the teacher feels is important. It is easier for children to daydream when they are free to move between areas and activities, as they can stop to ponder before moving on or returning to their investigation. If a child does not appear to be paying attention, it may be because he is internally processing thoughts, or that something she has just observed has triggered internal memories. For the thinking, internal processing and connections to be made there needs to be uninterrupted time within the 'untidy' classroom. Part of the chunking of time mentioned in the previous section involves tidying away resources at the end of a session. Given that for practical and health and safety reasons there has to be some control over the amount and organization of resources within the environment, we should still endeavour to enable children to retain their projects so that they can return to them at a later time, having processed their thoughts.

Key points

- **Teachers should allow some untidiness in the interest of children making connections.**
- **Children should be able to return to their projects at a later time to further pursue their enquiry in the light of internalization of thoughts, connections with previous experiences and new possibilities.**
- **Freedom to move between classroom areas and resources is necessary as part of the child's explorations and investigations.**
- **Daydreaming is a positive feature of learning.**

Conclusion

This chapter has examined some of the features of scientific exploration in young children and has made a clear case for the requirement of teacher knowledge and understanding of the nature of scientific enquiry in young children. For children to reach their scientific potential the teachers must recognize and cultivate their developing interest and skills.

Task 2.1

Valuing children as budding scientists

Track a child over a sustained period of time, for example a morning session. Look for evidence of involvement in scientific exploration. Use the table below to record your observations.

Table 2.1 Scientific explorations by budding scientists

OBSERVATION	EVIDENCE OF SCIENCE LEARNING
Where is the child?	
What are they doing?	

References

Adams, S., E. Alexander, M.J. Drummond and J. Moyles (2004) *Inside the Foundation Stage: Recreating the Reception Year*. London: ATL Publications.

Bruce, T. (1991) *Time to Play in Early Childhood Education*. London: Hodder & Stoughton.

Claxton, G. (2002) *Building Learning Power*. Bristol: TLO Publications.

Ellis, S. (2005) 'Mind Your Language'. *Nursery World* (24 February): 12–13.

Harding, S. (2005) Outdoor Play and the Pedagogic Garden, in J. Moyles (ed.) *The Excellence of Play* (2nd ed.). Maidenhead: Open University Press.

Haywood, H.C. (2004) Thinking in, Around and About the Curriculum: The Role of Cognitive Education. *International Journal of Disability, Development and Education* 51(3): 231–252.

Howe, A. and D. Davies (2005) Science and Play, in J. Moyles (ed.) *The Excellence of Play* (2nd ed.). Maidenhead: Open University Press.

Jenson, E. (1995) *The Learning Brain*. San Diego, CA: Turning Point Publishing.

May, P., E. Ashford and G. Bottle (2006) *Sound Beginnings: Learning and Development in the Early Years*. London: David Fulton.

Mueller, E.T and M.G. Dyer (1985) Daydreaming in Humans and Computers. *Proceedings of the Ninth International Joint Conference on Artificial Intelligence.* University of California, Los Angeles. August 18–24, 1985. Los Altos, CA: Morgan Kaufmann: 278–280 (retrieved from http://pdf.aminer.org/000/389/880/daydreaming_in_humans_and_computers.pdf).

Orlandi, K. (2012) *Onwards and Upwards: Supporting the Transition to Key Stage One.* London: Routledge.

Robbins, J. (2005) *Interweaving the Stitch and the Fabric: A Socio-Cultural Perspective on Researching Children's Thinking.* Paper presented at the British Educational Research Association Conference, Pontypridd, Wales (September).

Siraj-Blatchford, I., K. Sylva, B. Taggart, P. Samons, E. Melhuish and K. Elliot (2003) *The Effective Provision of Pre-school Education Project (EPPE) Technical Paper 10 Intensive Case Studies of Practice Across the Foundation Stage.* London: University of London, Institute of Education.

Waters, J. and S. Begley (2007) Supporting the Development of Risk-Taking Behaviours in the Early Years: An Exploratory Study. *Education 3–13* 35(3): 365–377.

3

The role of talk in developing scientific language in the early years

Babs Anderson

Introduction

The importance of the role of talk and discussion in developing scientific language cannot be underestimated. To a certain extent, the means of learning science is through the medium of increasingly specialized use of language and terminology, in the articulation of a precise meaning through the use of technical vocabulary. Wellington and Osborne (2001) suggest that one of the ways a teacher can support the development of scientific language in children is the practice of these terms, while for de Bóo (1999: 124) encouraging children to play with words helps them to take ownership of new vocabulary. Gelman and Brenneman (2004) stress the importance of valuing children's natural ability to learn new words relevant to specific science concepts and talk about not 'cheating' on vocabulary. Young children are ready imitators, so that given the opportunity to explore new vocabulary they will generally do so with pleasure. These technical terms need to correspond with their activity, taking into account their previous knowledge and understanding. When such terms are used in an appropriate context and the children can apply these in their discussions, they are able to exchange ideas about scientific concepts with their peers and adult carers.

Many of the activities in the early years classroom are opportunities for the children to learn the language of science and to develop their own ideas in a scientific way. However, this is not a simple task. The intention of the teacher to convey information (in the form of transmission) regarding a given concept may override the learner's need to internalize new information by action, discussion or space for thinking in order to accommodate new learning into existing schema (Piaget and Inhelder 1966). The revised early years foundation stage (DfE 2014) documentation, which guides practice in England, recognizes the primacy of language and communication in providing a rich environment of language, so that the children can express their ideas and thoughts, while listening to the thoughts and intentions of others. The use of language therefore to promote scientific thinking needs to include the development of the children's own ideas, including the evidence on which they are basing these ideas (Coltman 2008). A common refrain given by young children to a question is 'Just because . . .', with no explanation or reason following, thus indicating that although they may understand the core concept, they have difficulty in expressing their ideas and the basis for these. It is essential therefore that the adult models their own thinking aloud in order to show a means of expressing scientific ideas in a shared language experience.

Key points

- **Children need opportunity to play with new science vocabulary.**
- **Adults should think aloud and model how to express science ideas.**
- **The use of correct scientific language and vocabulary supports young children making sense of their world.**

Newton and Newton's (2001) work on the questions asked by primary school teachers illustrate the need for sound scientific subject knowledge on the part of teachers. They propose that the quality of questions asked is associated with the level of knowledge held by the teacher, so that those teachers with a science background asked more relevant questions and more 'cause and effect' type of questions. This must also be recognized in the early years forum, where a potential drawback is the belief that the practitioner's adult knowledge is sufficient to scaffold young children's development scientific conceptual framework, simply because it is adult.

The key characteristics of effective learning (DfE 2014) are useful to examine how children make sense of their world. The characteristic of engagement covers the aspect of playing and exploring, whereby the children are active agents in finding out about their environment, both physical and social, exploring by using their senses of smell, taste, touch, sight and hearing. They play with what they know and, in so doing, consolidate their understanding of new learning. A willingness to 'have a go' is vital, so that the child may learn from cause and effect. Their actions have a consequence, and being willing to partake in an activity enables the consequences of that action to unfold. Pushing over a tower of bricks may be an expression of a schematic behaviour (trajectory), identified by Piaget and Inhelder (1966). Yet the understanding of concepts afforded by this behaviour can be extended by the child's articulation of their prediction as to how the bricks would move and the reasons for their action. Skilful questioning by the teacher can ask the child to examine:

- Are some towers easier to push over than others?
- Why do they think this is the case?
- How could they make a more stable tower that will be harder to push over?

The dimension of motivation covers the aspect of active learning, which includes the children being proud of how they have achieved their intentions, their keeping on trying when they encounter a setback and their concentrating on the topic at hand. The dimension of thinking includes being creative and thinking critically. This involves the children having their own ideas and making connections to their previous experience in order to predict how a similar item may work or a person will behave. They are then able to apply their knowledge and understanding in order to choose how to carry out their actions. Careful use of scientific enquiry can guide children in their understanding of their world, making sense of their physical world and their community by exploration and observation.

How does the teacher help the child develop more precise language?

An adult can play the part of a facilitator of exploratory talk, where new concepts and vocabulary can be introduced by the adult without overt instruction (Figure 3.1).

Figure 3.1 Exploratory talk supports the open-ended verbal exploration of an enquiry.

Examples for the properties of materials include transparent, waterproof, rough, smooth, soft and hard and some of these terms are similar to those in everyday usage, whereas certain terms are more specific: for example, transparent. The children can use a torch to shine through materials, finding out which will block light or will let light travel through. They can then select a suitable fabric for a bedroom curtain. Using the correct term *transparent* here is important because 'see through' reinforces the misconception that the eye is active and that we see through a material rather than the material letting the light through.

How does the teacher support a child to refine their understanding of scientific concepts?

The teacher can ask the child to explain their thinking, so that they are accustomed to articulating their ideas without the adult dominating the learning trajectory. They may also use previous shared experiences to enhance the child's understanding, for example:

- Do you remember when we . . .?
- Do you think the same thing will happen here?
- If not, why not? What is different about this occasion?
- Does it work, when . . .?
- Why do you think that is?

This action of scaffolding requires child-centred teaching, so that the child's potential understanding is framed by the support of the questions of the more knowledgeable other (Corden 2000). Sensitive feedback is also essential, so that through this the practitioner can confirm if the children are succeeding in their ideas, or whether they need to suggest to the children that they consider an alternative if the current idea is proving problematic.

How does the teacher ascertain the developing scientific understanding of young children in order to plan for next steps or potential lines of development?

A record of the children's discussion enables an accurate assessment of their thinking to be made in order to inform future planning for science. An adult can record this in writing at the time: for example, they can annotate a digital photograph using the *Balloon Stickies* app (Figure 3.2) or use a digital voice recorder in order to reflect on this learning.

Recording enables the practitioners, as a team, to reflect on the children's discussions, giving a wider range of perspectives when making an assessment.

Figure 3.2 The *Balloon Stickies* app is a quick and easy way of recording children's talk.

Harlen et al. (2003: 58, 59) list core enquiry skills that young children may be encouraged to develop as part of the pedagogy of science:

- Observation
- Prediction
- Problem solving
- Decision making
- Communication.

Reasoning, comparison and linking cause and effect may also be added to the list, as these are associated with the use of talk to link prior experiences to the current activity.

Exploratory talk supports the open-ended verbal exploration of an idea or enquiry, so that the children learn to utilize the skills above in a natural context. Encouraging children to listen to the ideas of others and respond to these provides an indicator of how the child is using their own level of knowledge and understanding to understand the responses of others. In parallel with the development of enquiry skills, the practitioner may aim for a coordination of a whole-class response to a scientific enquiry by identifying and probing children's ideas, to refining them and guiding the class towards coherent scientific understanding (Kawalkar and Vijapurkar 2013). This process can support the practitioner in their analysis of the learning gains made by the class of children as well as individuals.

How can the teacher encourage the children to record their own ideas in a meaningful way?

The children can use mark making or drawing to record their ideas, but the use of technology is becoming increasingly valuable in the early years classroom. Digital photographs taken on a camera or tablet computer are useful to provide a pictorial record for the children to discuss after the event, where either the children or the adult can use the device to record the process. These devices utilize an accurate pictorial representation of the activity. These resources also enable the recording of the event, both visually and graphically, where either the adult can encourage the children to make their own labels for the sequence of photographs throughout the process or the adult can scribe for the children. The latter scribing function also enables the adult to use more precise terms when talking to the children about their activity, for example the use of the word 'push' rather than 'knock over' in order to connect with the idea of forces acting on objects.

Following children's interests is a key element of play-based pedagogies, as advocated in the *Statutory Framework for the Early Years Foundation Stage* (DfE 2014) and this can offer a valuable source of planned activity with the children. In the case study below, an example of this explores how an expressed interest was used as the stimulus for an experimental investigation.

Case study: Humpty Dumpty and his wall

The reception class theme was nursery rhymes, in order to familiarize the children with their cultural heritage. As part of the topic, the children learned a range of traditional rhymes, such as Jack and Jill; Hey Diddle Diddle; Mary, Mary Quite Contrary; The Grand Old Duke of York; and Humpty Dumpty. The classic illustration of Humpty Dumpty as an anthropomorphic egg – that is, an egg with the attributes or features of a human – particularly intrigued the children. One child had eaten a boiled egg with 'soldiers' – that is, long fingers of toast to dip in the egg – and thought it interesting that Humpty Dumpty had human features but looked like an egg, similar to the one he had eaten. Another child interjected that they had seen someone drop an egg and commented it was 'all sticky'. The two children developed a line of enquiry, wondering whether Humpty Dumpty was a boiled egg or not.

The teacher recognized the opportunity to develop their scientific thinking further, and asked 'How can we find out?' The children together with their teacher repeated the words of the poem:

> *Humpty Dumpty sat on a wall,*
> *Humpty Dumpty had a great fall,*
> *All the King's horses and all the King's men*
> *couldn't put Humpty together again.*

The teacher's careful questioning led the children to consider:

- *How and why did the egg fall?*
- *What was the height of the wall; that is, what made it a great fall?*

The children concluded they would need to build a wall and let the egg fall off, in order to replicate the accident, so then they would be able to observe what actually happens to an egg when it falls from a height. The teacher's input then reminded the children of their original question. This was whether Humpty Dumpty was a boiled egg or not. Wary of intervening in their ideas, yet understanding the potential for learning by sensitive adult questioning, the teacher suggested that perhaps one way of finding out was to let the same thing happen to both a boiled and a raw egg.

Thus the idea of a fair test could be integrated into the children's own scientific musings. The children readily agreed and two eggs were obtained, one boiled and one raw. Then the children took their learning outside, as the illustrations showed Humpty Dumpty to be sitting outside on a wall.

The rest of the class were invited to watch the results of the experiment. Two walls were built with the big wooden blocks and plastic sheets placed underneath, to contain the egg when it fell. The initial two children involved in the enquiry were asked by the teacher why they had built two walls of the same height, so that they could share their learning with the rest of the class. The terminology used by the children in their responses included the words 'fair', 'not fair', 'same' and 'different', as the children attempted to explain their understanding of how the height of the wall would have an impact on what happened to the egg. Thus they were grappling with the concept of gravity, recognizing that this force would 'pull' the egg downwards. Their understanding of this concept, however, appeared to outstretch their ability to fully articulate this understanding. Nevertheless, the beginning of the idea of a fair test is in the grasp of these children conceptually and is one that young children relate to easily. Young children can fruitfully ponder the following questions:

- *Is it fair?*
- *Do both need to be the same or can something be different?*

In this particular instance, the time required beforehand for these two children to be able to think through their ideas with a sensitive adult enabled them to process their own thinking. This in turn was crucial to their learning, as a quick and ready answer may not have led to the richness of the discussion and the openness to some adult input into their plan of action, with the children still retaining a sense of ownership of the plan.

The opportunity to observe closely and to determine differences was heightened by the sense of fun and excitement, especially when comparing the two eggs after their fall. Most of the watching children wanted to touch the raw egg in particular, developing a range of descriptive vocabulary, including 'sticky', as they attempted to explain the viscosity of the egg and how it differed from glue as another 'sticky' substance. The children were able to talk together through a range of shared experiences within the school setting in order to make connections with their prior understanding of 'stickiness', trying to decide whether the raw egg was like glue or wet paint or whether it was more like jelly that they liked to eat at birthday parties. The use of a digital camera also enabled a record of the experiment to be made, so that future discussions could relate back to the physical experience.

How can recounts develop scientific language?

The use of recounts is valuable, so that children are encouraged to think through their activity in a logical manner after the event – what they did and what they found out – by talking though the stages of their actions. For example, in the building of a tower, the child can be encouraged to enumerate the steps they took: first we collected all the big blocks together, then we put them together so that we built a tall tower. This understanding of sequence and classification is important as they are key concepts in the development of scientific knowledge and understanding. Other key concepts include identifying materials and differentiating between properties of materials. A recount of the play activity of washing a doll's clothes may include the concepts of waterproof and absorbent, when the child plays with items such as bibs, one with waterproof backing and one made of terry towelling.

The plan–do–review of High/Scope

High/Scope is an American programme that has influenced early years education internationally. This programme is based on the Piagetian principle that active learning is essential in children's learning and development (Hohmann and Weikart 2002). As part of this process, children engage in a plan–do–review activity, a three-part process in which language and communication play an integral part.

The planning stage involves the children stating what they are going to do. This can be in words or gestures. If the latter, the adult can put into words their interpretation of the child's intention. It is important that the adult understands, however, that this will be an interpretation of the child's intentions rather than an exact match. The use of the questions 'what will you do?' and 'how will you do it?' encourage a focus on the process rather than the question 'where are you going to work?' which encourages a focus on place. Adults are also able to support the planning process by discussing what might be barriers to their ideas.

The second stage is termed 'do' or 'work time', where the child carries out their intention, either alone or with peers. The role of the adult here is to discuss the child's activity, by responding to the child's conversation rather than to interrogate them and divert or subdue the child's own initiative. To do this successfully, Hohmann and Weikart (2002: 216–217) suggest that practitioners 'ask questions sparingly', 'relate questions to what the child is doing' and 'ask questions about the thought process'.

The third phase is review or recall. During recall the children construct a memory of what they have done and what they have found out. They are able to construct the mental representation of their activity. Over time, they are able to identify the links with their plan with the support of the practitioner, by discussion and conversation with regard to the shared experience.

Task 3.1

Thoughtful questioning by adults helps children to articulate their thinking and their use of scientific vocabulary. Observe how a child responds to questions asked and reflect on their use of scientific language as they talk about the activity. The table gives some possible starting points for questions. Use this to record how children respond.

The use of stories and storytelling, including traditional and classic stories

A wide range of picture books lend themselves to enquiry of a scientific nature, as factual picture books may be used to promote the acquisition of information, rather than the scientific process (Figure 3.3). The basic key process skills in science, according to Monhardt and Monhardt (2006) include raising questions, predicting, planning, observing, measuring, inferring, interpreting and communicating. These skills are easier for young children to comprehend and utilize

Table 3.1 Children's responses to questions

Questions	These can begin with:	How did the child/children respond?
Questions which encourage children to:		
• **talk about their observation**	What can you see?	
• **ask questions**	What does it feel like?	
	What would you like to ask?	
	What would you like to find out?	
• **talk about comparisons**	Is it the same as . . . ?	
	Have you seen anything like this before?	
• **talk about their predictions**	What do you think will happen if . . . ?	
	What do you think will happen next?	
• **talk about exploring and investigating**	How do you think you can find out?	
	When did it happen?	
• **explain**	What do you think is happening?	
	I wondered why	
	Why do you think . . . ?	
	What do you think is happening?	
	Why do you think this happens?	
Questions which encourage children to make links	When else have you seen this happening?	
	Do all . . . ?	
Questions that foster productive conflict by challenging the responses of one or several children	Where did the water come from?	

Figure 3.3 Narrative story provides a richly illustrated context for scientific enquiry.

in a context that is meaningful for them. The use of narrative story supports the children's understanding by providing a richly illustrated context for events.

In each of the examples below, a picture book is used to facilitate scientific enquiry, following the children's interests.

Ezra Jack Keats (1976) *The Snowy Day*. London: Puffin

This is the story of a boy exploring the nature of snow. This book is best used after the children have experienced snow for themselves so that they can consider what they have learned already about the nature of snow.

The teacher can record any questions asked by the children, so that a documentation of their interests is made explicit. The use of a spider diagram of what the children already know and think is a good record of their initial ideas. It is possible then to ask the children to predict what would happen to snow in different circumstances, for example if it were squashed into a snowball or warmed by hands. They should be encouraged to tell why they think this, linking to their prior experience and learning. A further question could be to ask the children whether we could make snow and compare this to making ice cubes in the freezer.

Once the discussion has come to a close, it is useful to revisit the spider diagram to articulate and model how knowledge is reinterpreted, by using another colour to illustrate knowledge that is either new or changed in the light of the group deliberations.

Eric Carle (2002) *The Very Hungry Caterpillar*. London: Puffin

This classic story tells of the metamorphosis of a caterpillar into a butterfly. This picture book is usually employed to support young children's understanding of the life cycle of butterflies. It also lends itself to questions such as:

- What kinds of food do animals eat?
- What are the differences between the adult and young of different species?
- How do caterpillars move in comparison to butterflies?
- What other animals lay eggs?
- What happens inside the chrysalis?

This is not a comprehensive list but records some of the possible ideas. The final question is one that lends itself to an appreciation of the awe and wonder afforded by science knowledge. It is a recognition that the process of metamorphosis is one of interest and challenge rather than just something that happens to produce a butterfly.

Pamela Allen (1988) *Who Sank the Boat?* London: Puffin

This book lends itself to the planning of an experiment. The topic of floating and sinking is a common one in early years settings and a water tray or sink is likely to be available as part of continuous provision, for the children to access without adult intervention. The book tells the story of a range of animals that decide to go out in a boat. The sizes of the animals vary: there is a cow, a donkey, a sheep, a pig and a tiny little mouse. The children can explore the idea of the boat changing from floating to sinking practically, using resources to work out their ideas.

Pamela Allen (1994) *Mr Archimedes' Bath*. London: Puffin

In a similar manner, this book tells the story of Mr Archimedes, who shares his bath with some animal friends. The displacement of water is key to the story, so that the bath overflows when additional bodies displace the water. Once they get out of the bath, the water returns to its original level (minus any water on the floor!).

Observing and measuring are possible responses to activity imitating the story line, so that as a range of objects are placed in a container of water, the children can see how this affects the level of the water (Figure 3.4).

David McKee (2012) *Not Now, Bernard*. London: Andersen Press

In one interpretation, this story book tells of a monster who eats Bernard. He is then mistaken for Bernard by his parents, who have ignored him throughout. The detailed illustrations are useful prompts to enable children to communicate their ideas as to why the parents think the monster is Bernard.

Traditional stories can also be used as a source of interest, for example:

- **Jack and the Beanstalk** to discuss the nature of plant growth
- **The Three Little Pigs** to examine materials
- **The Little Red Hen** to discuss the process of making bread
- **The Three Billy Goats Gruff** to plan how to build a bridge across the river.

Other stories lend themselves to the children developing their own questions. For example, Linda Atherton in Chapter 4 explains how one teacher used **The Gingerbread Man** as a starting point for thinking about dissolving and the properties of materials. The flexibility afforded to children's creativity by following possible lines of development (PLODs) instead of a predetermined plan is invaluable.

Figure 3.4 Teacher and children share *Mr Archimedes' Bath*.

The use of puppets in promoting exploratory talk

Puppets have long been used in literacy, PSHE, RE and drama within the primary school to engage children in articulating their ideas and exploring ideas, concepts and feelings (Figure 3.5). Simon et al. (2008) investigate whether puppets might also be usefully employed to promote talk in science. Their findings show how teacher participants were able to use the puppets to create a stimulating and engaging context, whereby the children were motivated by their interactions with the puppet. The use of puppets also increased the quality of scientific teacher talk, moving from recall to more reasoning and evidence-based assertions. The use of puppets would therefore appear to be a useful tool for scientific enquiry in addition to their more traditional uses.

Figure 3.5 Puppets can be usefully employed to promote talk in science.

References

Coltman, P. (2008) How Many Toes Has a Newt?: Science in the Early Years, in D. Whitebread and P. Coltman (eds) *Teaching and Learning in the Early Years* (3rd ed.). Abingdon: Routledge.

Corden, R. (2000) *Literacy and Learning through Talk: Strategies for the Primary Classroom*. Buckingham: Open University Press.

de Bóo, M. (1999) *Enquiring Children, Challenging Teaching*. Buckingham: Open University Press.

DfE (Department of Education) (2014) *The Early Years Foundation Stage*. London: DfE.

Gelman, R. and K. Brenneman (2004) Science Learning Pathways for Young Children. *Early Childhood Research Quarterly* 19: 150–158.

Harlen, W., C. Macro, K. Reed and M. Schilling (2003) *Making Progress in Primary Science*. London: Routledge.

Hohmann, M. and D. Weikart (2002) *Educating Young Children: Active Learning Practices for Preschool and Child Care Programs* (2nd ed.). Ypsilanti, MI: HighScope Press.

Kawalkar, A. and J. Vijapurkar (2013) Scaffolding Science Talk: The Role of Teachers' Questions in the Enquiry Classroom. *International Journal of Science Education* 35(12): 2004–2027.

Monhardt, L. and R. Monhardt (2006) Creating a Context for the Learning of Science Process Skills through Picture Books. *Early Childhood Education Journal* 34(1): 67–71.

Newton, D.P. and L.D. Newton (2001) Subject Content Knowledge and Teacher Talk in the Primary Science Classroom. *European Journal of Teacher Education* 24(3): 369–379.

Piaget, J. and B. Inhelder (1966) *The Psychology of the Child*. London: Routledge & Kegan Paul.

Simon, S., S. Naylor, B. Keogh, J. Maloney and B. Downing (2008) Puppets Promoting Engagement and Talk in Science. *International Journal of Science Education* 30(9): 1229–1248.

Wellington, G. and J. Osborne (2001) *Language and Literacy in Science Education*. Buckingham: Open University Press.

Resources

The research project Creative Little Scientists is a collaborative European research project. Part of its remit is the production of a wide range of deliverables, which support a creative pedagogy in science and mathematics early years education. These deliverables are a rich source of examples and illustrations of how to use exploratory language to support understanding.

Creative Little Scientists Consortium (2012) *Creative Little Scientists: Enabling Creativity though Science and Maths in Preschool and First Years of Primary Education*. Retrieved from: www.creative-little-scientists.eu/content/deliverables.

4

Starting points to inspire science in the early years

Linda Atherton

Somewhere, something incredible is waiting to be known.

Carl Sagan

Young children need to make sense of their world and the world around them. According to the *Statutory Framework for Early Years Foundation Stage* (DfE 2014) there are three characteristics of effective teaching and learning: first, playing and exploring, where children will start to investigate and experience things and 'have a go'; second, active learning, which means that when difficulties arise they persevere and subsequently enjoy their achievement; finally, creating and thinking critically to develop strategies for doing things.

In order for these to happen the adult needs to think carefully about where the children are starting from. What do they already know? If the adult is clear about this then it is possible to plan appropriate opportunities for learning to take place. Careful observations and skilful questioning can do this and so recognize prior learning and experience, thus identifying how to proceed. In her review of the evidence for the early years foundation stage, Tickell (2011: 4.3) says we should be paying attention to what children enjoy and how they respond to different things, then using this knowledge to provide an enjoyable and stimulating environment that helps to extend children's development and learning.

This chapter will explore how important starting points are in setting the scene for learning and subsequently establish what will engage and allow learning to take place. The imperative of making starting points real and relevant to the child, providing an incentive for exploring scientifically and possibly finding an answer to a question or solving a problem, will be discussed. Emphasis will be placed on the importance of a 'hook' to catch their attention and pull them into a tangible activity (Figure 4.1). This stimulus for an activity or investigation in science needs to encourage a genuine interest and generate curiosity, which will

Figure 4.1 Here the snail is a 'hook' to catch the children's attention and generate curiosity.

lead to purposeful play and exploration. We should never underestimate the importance that fun and the 'wow' factor at the start can have on a child's experience.

This does not mean that everything that is given to a child must be new and innovative. Practitioners must not feel that they have to keep inventing new ideas and experiences to give to children. It is important to look afresh at everyday resources, consider their potential and find the extraordinary in the ordinary (Brunton and Thornton 2010: 17). This chapter uses worked examples with readily available materials and resources to show how this could be realized in an early years setting.

Macintyre (2001) wrote about the importance of exploratory play. However, successful learning only takes place when the child is encouraged to take risks and be open to new problems in a secure environment. Successful teachers therefore need to offer challenges in thinking that, while not posing a threat to their security, encourage children spontaneously to begin internalizing questioning and hypothesizing to provide scaffolding for themselves (Adey and Shayer 1994). This was neatly summed up as 'comfortable challenge' by Merry (1998: 116–117). A characteristic of effective learners is that they ask questions of themselves and others (Fisher 1995). Consequently, early years settings should develop the climate and culture for this to take place in. Providing concrete experiences for young children which can allow, eventually, the ability to deal with abstract ideas is essential to develop this culture. The hooks then need to be tangible and real.

Katz (2010) notes that young children have an in-born disposition to learn. Moreover, research has shown that introducing too formal a curriculum at too early an age may have short-terms gains in skills and knowledge but in the long term tends to damage their disposition to be learners. So it is important to concentrate on what children are doing rather than the end product. Once again this emphasizes that young children need to have opportunities for exploration and play. Thus child educators will find it difficult if not impossible to provide a high-quality curriculum without providing high-quality play and so 'infect the children with enthusiasm' (Wood and Attfield 2005).

Moyles (1989) talks of a play spiral, where free play leads onto directed play and the resulting free play leads once again to directed play, until the child is able to restructure the events to make them meaningful. The skill of the practitioner is to orchestrate the learning by providing a stimulating climate where productive questions can be asked (see Chapter 5) and there is time to explore and discover. Yet the notion of children simply spontaneously discovering scientific principles has been discredited (Harlen 2000). Thus planning for open-ended investigation where discovery can occur needs to have careful scaffolding of scientific play where practitioners have a good scientific understanding. Accordingly exploration, both child initiated and adult initiated, can be guided with deliberate scaffolding of scientific concepts (Davies and Howe 2003).

Throughout this process of play and exploration the skills of young scientists can be developed. Actual starting points are important in this process to ensure authenticity. Children should not be asked to *pretend* to be a scientist; the expectation should always be of *being* a scientist. Value should be placed on acting in an appropriate manner because this is one way that children can think about their world (Beauchamp 2013: 288). Beauchamp highlights the need for subject rigour to ensure correct science knowledge, so that, when appropriate, children view their learning through a 'science lens'. The scientific communication, social and physical skills, and in particular fine motor skills, need to be nurtured. Developing scientific attitudes and dispositions is also important (Brunton and Thornton 2010: 14). All these raise confidence and self-worth.

As previously mentioned our task is to 'infect the child with enthusiasm' (Wood and Attfield 2005) with 'open-ended' activities that can be accessed by different ability levels. The teacher's role is to place a pedagogical framing to this. How do we start this process?

Key points

- Starting points need to be authentic.
- Find the extraordinary in the ordinary. The initial hook does not have to be new or innovative.
- Successful teachers challenge children's thinking, while not posing a threat to their security.

Creating the right environment

We have to recognize that learning cannot be separated from experience and that learning is related to relationships and personal interests. Emotion and feelings also have a vital role to play in what we might later come to identify as intellectual learning. The implication of this is that the way an activity is introduced and the culture and climate the teacher has constructed will affect the learning process. Reggio Emilia schools in Italy give careful attention to the look and feel of the environment, which is often referred to as the 'third teacher'. Moreover, Moyles (1989) felt that by providing play and situations in which both physical and mental skills can be practised would lead to more challenging play. She also advocated using different terminology for a science activity, so instead of saying 'play in water' practitioners should be thinking about 'working with water'.

For scientific enquiry, then, careful thought needs to be given to the environment. The area must motivate and challenge children; this must be more than a nature table or science display. The challenge is to provide an enriched environment in which children play and learn and have first-hand experience of science phenomena. Thinking carefully about the space in a classroom or outdoors is important. The more cluttered and disorganized the area the less the likelihood is of introducing an element of surprise and a new starting point (Smidt 2002). The right tools need to be provided. The addition of hand lenses for looking carefully and tools for measuring immediately makes a display more interactive. Simple cue cards in an indoor or outdoor space help with guiding children. These cue cards, the size of postcards, can be laminated and have clear visual/graphic/picture instructions such as an eye, to encourage children to look carefully. An arrow encourages children to look in different directions including up and down, an ear encourages them to listen and a hand encourages touch. Teachers might decide that they want to focus on a particular sense so a sensory walk can be made by placing just one or two of the cards in an environment. This initially focuses the child's thinking. Once they have become familiar with this activity children can then, with the aid of an adult, set up new learning walks for other children to follow.

Case study

An early years teacher identified the need for her children to have an area where they could play at being scientists. The children decided to call this a science lab. The teacher carefully set up the area with scientific equipment because she wanted them to use proper science equipment that they would encounter later on in the school. When asked about the science lab area she wrote:

> *The science lab – they loved dressing up in coats and goggles etc. and using clipboards and science equipment. We had parents in to share how they work there and to play alongside the children. The children like 'experiments' with as much colour and fizz as possible. They like things regularly changed so that it is exciting. They like plastic bottles and writing labels and test liquids. They loved practical equipment and hands-on experience – they could take the lead and work with another child to share ideas well. (Looking for the potion to . . . Help them fly etc. – H&S never drink!)*
>
> *Official-looking paper to write on, sticky labels, name tags, charts and observations were useful. Posters – long words! – things hanging from the ceiling – information books – signing in, all helped.*

To keep them engaged I always find including them in planning the next step/asking questions to be answered by their work and just taking them seriously.

It is interesting that she felt that being taken seriously was important. They were being real scientists and spent a lot of time making the area authentic. Some of the children invited older children to come and investigate in their classroom. They were then invited into Key Stage 1 to show and tell other children what they had found out about mixing and making potions.

Starting points in everyday science

The beginning is the most important part of the work.

Plato

The challenge then is to infect the children with enthusiasm so purposeful play and exploration can take place. Open-ended activities can make this happen. There needs to be an emphasis on the link between high-quality play and a high-quality curriculum because, while some opportunities for learning can be carefully planned, others may be spontaneous and unexpected.

Simple science equipment can provide a starting point for science. In a local nursery setting a teacher had left some pipettes on a table near a bowl of water

(Figure 4.2). She left the children to explore these. At first they were unsure what to do. Through exploration and trial and error the children discovered that they could squirt water and make drops. As they practised they became more confident and adept at using them.

What resulted from this activity was that children started to investigate using pipettes for other liquids nearby. The children were fascinated by the fact that some liquids were much more difficult to suck up. At this point an adult intervened by taking the children outside where a large piece of lining paper had been put on the floor. She then simply asked 'I wonder what will happen if we squirt our liquids onto the paper?' What ensued were children at first just squirting liquids. However, they started to become more focused and the activity

Figure 4.2 A pipette was used as a starting point.

became the starting point for developing science vocabulary. At this point the teacher introduced scientific vocabulary such as 'viscosity' and the children enjoyed hearing the words and practised saying them. It is worth noting that the teacher recognized that the children's learning would be supported by using the correct science vocabulary, as discussed by Lois Kelly in Chapter 1 and Babs Anderson in Chapter 3. This foundation could be built on by careful intervention and the practitioners providing the correct sources of a rich environment for learning. The reason that this was such a dynamic activity was that the teacher had used a 'scientific lens' and carefully planned what resources might be needed for the activity. She had started from a very simple starting point and by allowing the children time to explore she had then gone through a spiral of understanding.

Observation is a key skill for young children (Figure 4.3) and a simple old-fashioned overhead projector (OHP) or a visualizer can be a fantastic starting point for observation. They provide opportunities for children to be able to engage with and examine ordinary materials from lots of different angles: for example, cogs, wheels and springs that are quite small can be viewed in more detail. An image can be projected onto a wall and seen by a group of children who can work together to talk about what they can see. This supports their language development and scientific thinking. The adult can also encourage children to focus their attention on different shapes or textures, or to compare shadows.

A digital microscope was a starting point for another group of children. The initial exploration was of how it worked and one of the boys, when it was his turn to use it, stuck it in his ear. With shrieks of excitement he realized that there was something in his ear. The teacher told him it was wax and with her help he captured a picture of it, as he wanted to show his friends. Noah then said that he had grommets and wanted to look down his ear to see if he could see them. The teacher helped him do this. While he could not see the grommet, his ear was full of wax. This then became an exploration by a group of children as to who had wax and who had the most. The initial starting point then had led to child-initiated exploration and enquiries. How had the wax got in there? Was everyone the same? This perhaps was not one of the possible lines of development (PLOD) that the teacher had expected but the children had become enthralled by possibilities. This led to rich learning experiences because it was real and relevant to the children. This shows that some starting points can be planned while others need to be recognized by the adult as opportunities for children to practise real-life skills without the presence or intervention of adults. In this situation, the successful teacher offered a challenge to the child's thinking and encouraged the children spontaneously to take on roles for themselves for scaffolding learning for other children. In Chapter 8, Eleanor Hoskins considers the role of technology to support children's observations.

Figure 4.3 Observation is a key skill for young children.

The power of stories

Stories are another source of starting points for science and are a very effective way to promote scientific thought, linking together other elements of the curriculum. Well-known traditional tales such as 'The Gingerbread Man' are ideal. In an early years setting, a group of children were questioned about what would happen to the gingerbread man if he were placed in water; they were very unsure. However, just by placing a gingerbread man in a glass of water and watching, the children could see the man disintegrating and instantly understand what was happening. From this simple starting point came questions about what would happen in hot and cold water or if it had to be water. The children were then encouraged to think of ways to get the gingerbread man over the river without his getting wet or being eaten by the fox. This simple problem solving activity led to many different possible solutions such as tying him to a rocket balloon, launching him in a catapult or attaching him to a helium balloon. Throughout the process the children were actively communicating to each other, critically looking at strategies and extremely focused as they became engaged with wanting to save the gingerbread man. Further examples of stories which can be starting points for early science learning are discussed by Babs Anderson in Chapter 3.

Ordinary materials can also inspire young children to think scientifically and are great starting points for science investigations. The advantage of these is that they can be easily sourced, making them ideal for education settings. Something that is readily available is a packing substance called packing peanuts. This is a material that is now generally used to pack fragile equipment coming in to schools. It should not be confused with polystyrene chips, which have been previously used by companies. The reason that polystyrene is not used is because if young children are exposed to small fragments there is a risk of ingestion and inhalation. If this occurs it does not show in X-rays so the material is very hazardous to use. Thus the packing industry has developed a safer product that is made from potato starch, and this dissolves in water. This still has the property of being able to protect fragile or delicate materials. However, it is much safer as it dissolves into non-allergenic materials. It is cheap and can be obtained from lots of suppliers on the Internet in 1 kg packs, which contain hundreds of pieces. This would be enough for many investigations.

Packing peanuts are a great resource for science. For young children the property of dissolving is a fantastic starting point. This action happens quickly and is very visual. Some teachers, when using packing peanuts, have renamed them 'magic corn' because of the speed with which they react with water and appear to disappear, usually in less than 30 seconds. One group of teachers in an early learning setting were reluctant to use the term 'magic' because this might detract from being scientific and might lead to misconceptions. However, a teacher in an alternative setting who had used the term stated that the children were really excited initially by the fact it was magic. Yet later on, through exploration, they realized that the 'magic corn' had to get wet to disappear. Noah said 'I know how the magic works – it needs water.' Noah and group of his friends then asked if they could find out what happens when other materials were placed in water, which led to further investigations.

In another setting 'magic corn' linked to a story was an effective starting point. As you read this account try mapping out some PLODs other than those described. One particular story that works well in early years is about a group of river animals (otter, frog, swan, water rat etc. – these can be represented by pictures, toys or puppets) who wanted to become the king's assistant. They argued among themselves so were set a challenge to carry the corn across the river. All they had to do was choose a material to wrap the corn in to keep it safe in the water and give it to the king. When a group of 4-year-olds did this Sophie declared 'I think the towel will win.' When asked by the teacher why she thought this, Sophie explained that it was because you use towels to get dry, so towels would keep the corn dry. This is a common misconception in young children. The corn was wrapped in a small piece of towelling, which was easily kept in place by elastic bands. The package was then put in a tray with water to simulate the river and a 30-second sand timer was used. After this time the towel was removed to shrieks of 'Oh no', 'It's snot' and 'Gone'. The magic corn had dissolved and was just a slimy residue on the towel. The speed that this happened had Sophie and the other children hooked. They all gathered round, amazed that it had vanished. This then stimulated lots of talk and questions about what might happen with other materials. The role of the adult in this situation was key in ensuring that there were different materials available so that further exploration could occur. This stimulated further discussion by the children as they compared what had happened to their magic corn when it was wrapped in different materials. Once again the activity was a starting point for the development of scientific language to describe the materials.

The initial hook of the corn led to an extension of this activity by looking at material under a simple digital microscope, which enabled the children to observe their materials from a different perspective. Sophie was amazed by how 'holey' the towel was and through careful questioning started to think about how the water had got through the holes. This helped to dispel her misconception of the towel keeping the corn dry.

Balloons are another readily available resource which can stimulate heuristic play. According to the *Chambers Dictionary* (2002), 'heuristic' means helping to find out or discover; proceeding by trial and error. It stems from the same root as Eureka – 'I found it!'

Balloons can be used as a starting point for exploring and investigating static electricity. A tray of different sized shapes and colours of tissue paper fish shapes is needed. A quick demonstration of rubbing the balloon and then placing it in the tray starts lots of thinking. How many fish can be picked up? Are big fish easier to pick up than small fish? Does the number of rubs affect how many fish I can pick up?

Many early years practitioners will be familiar with ice balloons, made by freezing balloons filled with water, as a starting point. They provide a rich learning experience for developing observation skills. Using a torch, children can talk about what they see as they shine the torch on the ice balloon and this can be extended by adding a few drops of food colouring and watching it seep into the ice. Once the ice balloon has been out of the freezer for a while and is beginning to melt the children can touch it and talk about how it feels. Sprinkling the ice balloon with salt can lead to further discussion about the sound ice makes as it melts. Working with a group of early years practitioners I developed this into a pass-the-parcel activity, which encouraged children to observe, make predictions and ask questions. An ice balloon was wrapped in layers of newspaper, clingfilm and foil and placed in a box and then wrapped up in a bigger box covered in brightly coloured paper. The children handled the parcel and discussed what might be inside. Through careful questioning, the teacher encouraged the children to use all their senses as they made their predictions. They explored how heavy it was and whether they could hear any sound coming from the box. After a while the children decided to unwrap the parcel. Inside was a box – once again, the teacher pursued ideas of what might be inside. Throughout this experience children were communicating their ideas. This was repeated with the next layer, foil, then the newspaper and clingfilm, which were all handled and investigated. Eventually the final layer was removed, revealing an ice balloon that could be explored (Figure 4.4).

One practitioner said:

> This was a great activity; it really got the children talking and getting more excited as the layers were removed. When they saw the balloon there was real wonderment. Immediately one child got a hand lens to look at it. We kept the ice balloon all day and the children kept going back to explore how it changed throughout the day. A fantastic way to get my children talking and thinking like scientists.

A fantastic starting point for sound is to place a hexnut in a deflated balloon which is then inflated. The challenge is to make the hexnut move inside the balloon. This is an easy manipulative skill for children and through trial and error they will get the hexnut spinning. It will 'sing' as the edges of the nut rub against the balloon.

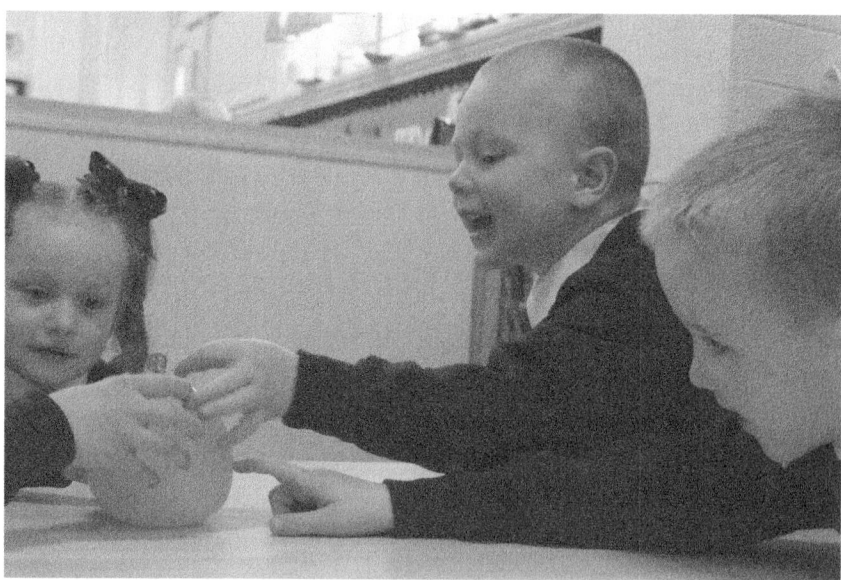

Figure 4.4 Look what we've found!

Children can explore how to change the sound and this can lead on to exploring the sounds made when different objects are placed inside balloons.

A simple chemical reaction can be explored using balloons. A teaspoon of bicarbonate of soda and half a teaspoon of citric acid are placed in a deflated balloon and the children feel the powders inside the balloon. The adult can then inject water into the balloon, using a syringe and quickly tie a knot (this needs to be practised!). The balloon inflates rapidly and can be passed around. The children can see the balloon inflate, hear the bubbles and fizz in the balloon and discuss the changes that occur. The reaction between citric acid and bicarbonate of soda, when dissolved in water, produces carbon dioxide and is endothermic so the balloon gets quite cold. I leave you to think about what might develop from this starting point.

Task 4.1

Possible lines of development (PLODs)

Use the observations in the grid below. In each case, consider what the possible lines of development for science could be.

Name	Observation/interests	Possible lines of development
Alfie	Shows great interest in snails. He spends a lot of time looking for them in the school garden	
Florence	Every day this week has headed straight for the coloured glasses and wears them for the whole session	
Isla	Parent reports that often talks about the stethoscope following a visit to the baby clinic with a new brother	
Jack	Has shown great interest in the construction work taking place in the building site next to the nursery	
Evie	Spends time tending the garden with a small trowel	
Nathan	Following our 'magic corn' activity has become fascinated by the digital microscope	

Conclusion

Where children start is important. The initial hook does not have to be new or innovative; it can be looking at the familiar with a much more enquiring mind. However, the teacher does need to understand not only what the child already knows but also what will motivate and challenge them to want to enquire. The role of the adult is to plan initial activities in a directed fashion that will result in the spiral of discovery recognizing how new opportunities can be introduced to enable high-quality and purposeful play. The challenge is to build learning bridges between what children already know and the possibilities for discovery of new knowledge.

References

Adey, P. and M. Shayer (1994) *Really Raising Standards: Cognitive Intervention and Academic Achievement.* London: Routledge.

Beauchamp, G. (2013) Understanding the World, in I. Palaiologou (ed.) *The Early Years Foundation Stage: Theory and Practice* (2nd ed.). London: Sage.

Brunton, P. and L. Thornton (2010) *Science in the Early Years: Building Firm Foundations from Birth to Five.* London: Sage.

Chambers Dictionary (2002) Edinburgh: Chambers Harrap.

Davies, D. and A. Howe (2003) *Teaching Science, Design and Technology in the Early Years.* London: David Fulton.

DfE (Department for Education) (2014) *Statutory Framework for the Early Years Foundation Stage* (retrieved from www.gov.uk/government/uploads/system/uploads/attachment_data/file/299391/DFE-00337-2014.pdf).

Fisher, R. (1995) *Teaching Children to Learn.* Cheltenham: Nelson Thornes.

Harlen, W. (2000) *The Teaching of Science in Primary Schools* (3rd ed.). London: Paul Chapman.

Katz, L.G. (2010) STEM in the Early Years, in *Collected Papers from the SEED Conference* (retrieved from http://ecrp.uiuc.edu/beyond/seed/katz.html).

Macintyre, C. (2001) *Enhancing Learning Through Play: A Developmental Perspective for Early Years.* London: David Fulton.

Merry, R. (1998) *Successful Children, Successful Teaching.* Buckingham: Open University Press.

Moyles, J.R. (1989) *Just Playing: The Role and Status of Play in Early Childhood Education.* Buckingham: Open University Press.

Smidt, S. (2002) *A Guide to Early Years Practice* (2nd ed.). Abingdon: Routledge

Tickell, C. (2011) *The Early Years Foundation Review: Report on the evidence* (retrieved from www.gov.uk/government/uploads/system/uploads/attachment_data/file/184839/DFE-00178-2011.pdf).

Wood, E. and J. Attfield (2005) *Play Learning and the Early Childhood Curriculum* (2nd ed.). London: Paul Chapman.

5

How can continuous provision inspire early years science?
Faith Fletcher and Di Stead

Introduction

Children are naturally curious about the world and want to make sense of their living environment. They want to know why bubbles are always round, why they have a shadow sometimes, how rain gets into the sky or who pulled the plug out at the seaside. But they do not want adults to give them the answers. They want to be the discoverers, the experimenters and the theory builders. They do not want science to be something that is imparted to them; they want it to be something that they do. They want to be scientists, not just consumers of science. They want to ask their own questions, collect their own data and arrive at new and wonderful ideas. These 'wants' can and should be achievable through well-resourced and planned continuous provision.

Case study 5.1

The home corner

Bradley (3 years), Suzanne (3 years 6 months) and Sophie (3 years 2 months)

The children were setting up the home corner. They had drawn a plan of what it should look like and had made a list of the furniture and white goods they required. Some equipment we had but not a washing machine; they made one from a large box. As the week progressed I would add items that the children either requested or that I believed would enhance discussion and play. In the second week I added a laundry basket. The children filled it with laundry (table cloth, tea towels and baby clothes)!

A long discussion ensued over the box washing machine and the fact that it wasn't up for the job; the laundrette was too far away. The decision was made to hand wash the items (Figure 5.1). This was then followed by a request for a washing line outside. This we did and the children pegged the clothes out to dry

Figure 5.1 Suzanne is concentrating and totally absorbed in the important task of washing.

in the wind. The children monitored whether they were dry. There were many discussions on how to hang them out and what sort of things would affect the drying time!

Role play is a powerful part of continuous provision and is a natural place for science to sit. In the role play in Case study 5.1, there is a good link between the outdoor and indoors. It began with the children's home experience brought into the setting. The children in this case study already understood what was required for a home to run successfully. The routine of laundry – washing, drying – is part of their lives (Figure 5.2). Role play provides a vehicle for talk and the role of the adult extends the talk by helping them reflect on the activity: for example, these children initiated a discussion about whose job it was to do the washing. Their discussion around 'when is a good drying day?' began their learning journey towards finding out about factors that affect evaporation.

Figure 5.2 Bradley demonstrates powerful fine manipulative control as he pegs out the washing to dry.

This chapter will consider the powerful role that continuous provision plays in the development of children's natural scientific inquisitiveness. Providing an 'enabling environment', we ensure that children have the opportunity to engage with scientific experiences, through thinking and talking. Children are provided with the environmental tools to allow them to follow their own interests at their own level and this enables them to articulate their ideas and the conclusions of their learning in an unhurried manner. This then ensures sustained shared thinking and opportunities for revisiting the same activities, which reinforces and consolidates their learning. It is acknowledged that children need opportunities to review ideas but also need to be introduced to new ideas.

What is continuous provision?

Continuous provision is defined as 'carefully chosen and organised quality resources placed in areas which are *always* available for children to access independently across every area of learning' and enhanced provision as 'the resources that are added to Continuous Provision within the indoor and outdoor learning environment which match the topic and the interests of children' (Birth to Five Service n.d.). David Hawkins (2002: 65) says that inherent in this idea is that 'it is the teacher who carefully prepares the environment, offering the children materials and equipment with which to engage. The teacher's preparation of the environment is based on her knowledge of the children's backgrounds and interests, combined with an understanding of children's learning, motivation, and development.' Alistair Bryce-Clegg, however, is concerned about the misconception that continuous provision is about the resources you have left out all the time. He suggests that the purpose of continuous provision is to 'continue the provision for learning in the absence of an adult' (2013: 4). It is not always about letting the child loose simply to play with familiar resources but it enables adults to draw them into experiences that they may not naturally go to.

The benefits of continuous provision are that it allows time for children to pursue their particular interests, at their own chosen rate and for a sustained period of time. Good continuous provision:

- allows children to follow careful observation
- takes account of children's predictable interests, allowing time for them to pursue them
- provides security and continuity for children's learning as part of an environment that allows them to take calculated risks
- provides opportunities for all areas of learning and development to be accessed, possibly permitting a chance to 'wallow' in ideas
- enables children to develop concentration and perseverance, revisit, examine and retest.

Enhancing continuous provision extends children's experiences when it:

- is linked to the theme or area of learning and development
- is linked to the interests of the children
- gradually adds interest to the continuous provision
- offers new possibilities of learning experiences through play.

In the light of experience, continuous provision is much more than both of these ideas. As a result of careful observation and knowledge of the developmental needs of the children the teacher provides resources, time and space that reflect children's interests and needs and provides opportunities for enhanced learning. It should also reflect the breadth, range and balance in terms of the seven areas of learning and development and children's interests. It should include indoor and outdoor environments. Those resources and experiences that are constantly available every day allow children to return to the same activity, to test or revisit theory (as discussed in Chapter 1) and they provide opportunity for trial and error. Continuous provision is very useful for learning about children's interests, which subsequently informs planning. Practitioners need to take the time to observe what preoccupies a child: for example, a mirror or fascination when turning a light switch or water tap on and off. They fiddle to see what happens, and notice the repercussions of their actions: for example, the child who sets the fire alarm off!

Continuous provision supporting science learning

For children in the early years, science is about exploring and enquiring in order to make sense of the world around them. They are sense makers, naturally curious, interested in their immediate environment; they are creative, imaginative; theorizers, collaborators and educators. Early years professionals are aware that children are competent theorizers: Each child has his or her own *baggage* of hypotheses on the possible sense and meaning of things. These hypotheses derive from children's personal experiences and they want to communicate them to others, adults as well as children' (Rinaldi 2006: 130).

Continuous provision provides opportunities for children to develop as young scientists through playing and exploring, being a creative and an active learner, and to develop the skill of being a critical thinker, capable of planning, assessing and adjusting their thinking and play. David Hawkins (2002: 68), in his essay 'Messing about in Science', speaks of time devoted to unguided exploratory work, when children can 'construct, test, probe, and experiment without superimposed questions or instructions' (Hall 2010). In addition, there are opportunities to engage with other people and their environment as they construct meaning. While the characteristics of effective learning (DfE 2014: 6) underpin learning and development across all areas and support the child to continue to be a motivated and interested learner, we argue that this is particularly relevant for learning science because these characteristics underpin ways in which children construct their own ideas about how the world works. These are:

- **Play and explore**
  - Find out and explore
  - Play with what they know
  - Be willing to 'have a go'
- **Be active learners**
  - Be involved and concentrate
  - Keep trying
  - Enjoy and achieve what they set out to do
- **Be creative and think critically**
  - Have their own ideas
  - Make links between the range of experiences which explore a specific science idea
  - Choose ways to do things.

These characteristics support and guide the area of learning and development 'Understanding the World', the aspects of which are easily explored through science in the continuous provision. A child achieving these characteristics will truly be a 'budding' scientist, as explored in Chapter 2. They will develop all the strategies and skills that are required to apply knowledge to make sense of their world. 'Science education is a process of conceptual change in which children reorganize their existing knowledge in order to understand concepts and processes completely' (Havu-Nuutinen 2005: 259). The word *process* implies something that happens over time with repeated encounters. These characteristics can also provide a useful framework when planning and evaluating experiences (see Chapter 9).

Why is play so important in learning science?

The importance of play in science cannot be overstated. Many research scientists will talk about how they play with ideas. Richard Feynman once said of his work, 'Why did I enjoy doing it [physics]? I used to play with it. I used to do whatever I felt like doing . . . [depending on] whether it was interesting and amusing for me to play with' (Feynman 1997: 48). Froebel (1887: 55) insists that 'play is the highest phase of child development of human development at this period: for it is self-active representation of the inner representation of the inner from inner necessity and impulse.' The main benefits of play are that it:

- promotes independence, confidence and self-identity
- enables children to consolidate and develop skills and concepts
- gives self-empowerment to the decision maker

- encourages exploration and enquiry
- allows children to create and enjoy elements of surprise, awe and wonder.

When thinking about how play supports children's science learning it is useful to consider how the 12 features of play identified by Tina Bruce (2005) underpin the natural emergence of children's enquiries into how their world operates, functions and interacts:

1   Children use first-hand experiences from life.
2   Children make up rules as they play in order to keep control.
3   Children symbolically represent as they play, making and adapting play props.
4   Children choose to play – they cannot be made to play.
5   Children rehearse their future in their role play. Dan Davies (2011) explores how children learn to be scientists through play; see also Chapter 6.
6   Children sometimes play alone.
7   Children pretend when they play.
8   Children play with adults and other children cooperatively in pairs or groups.
9   Children have a personal play agenda, which may or may not be shared.
10 Children are deeply involved and difficult to distract from their deep learning as they *wallow* in their play and learning.
11 Children try out their most recently acquired skills and competences, as if celebrating what they know.
12 Children coordinate ideas and feelings and make sense of relationships with their families, friends and cultures.

Task 5.1

Part 1

Look at these photographs of children playing in continuous provision. Beside each photograph are some observations that the teacher made. Can you identify which of Tina Bruce's 12 features of play are illustrated?

Part 2

- What science ideas are the children developing?
- What are possible lines of development?
- How might you enhance this provision?

Sonia - Date: 25th April Time: 10.00

What is observed?

Sonia is involved with solitary play in the dry sand (Figure 5.3). She is using a scoop in her right hand to pour sand into a coffee pot. Her grasp is strong and she is concentrating hard.

Evaluation

Sonia understands that the scoop needs to be at a high level in order for the sand to fall. Her hand–eye coordination is developing.

Next steps

Mastery of pouring comes with experience. Sonia needs the opportunity to continue to revisit the experience as many times as necessary and the opportunity to pour a whole variety of different materials, e.g. rice or bubble bath.

Figure 5.3 Sonia playing alone.

Jemma - Date: 2nd April Time: 11.00

What is observed?

Jemma is playing with a friend in a trough which has stones and water in (Figure 5.4). She is experimenting with stones hitting the water then stones hitting stones. She repeats this over and over again and then tries dropping the stones from different heights.

Evaluation

Jemma already has a good idea of what is about to happen. Experience has told her that dropping the stone on the stones rather than the water will make a louder noise. Her eyes are tight shut in anticipation and she is bracing herself for the loud noise.

Next steps

Fill the water trough with lots of water.

Fill the trough with sand.

How does the sound change?

Figure 5.4 Jemma dropping stones.

Matilda - Date June 17th Time: 2.30

What is observed?

Matilda has rolled up her sleeves and is clutching the jelly-type material (Gelli Baff) in both fists, opening and closing them (Figure 5.5). She is sharing the experience with two other children.

Evaluation

Matilda is obviously enjoying the activity, not only handling the Gelli Baff but also the social aspect.

Next steps

Talk with Matilda about the colour and the texture. Is it a solid? What might it be used for? What will it be tomorrow?

Next time mix more water with the Gelli Baff to make it runny.

Add a range of tools such as jelly moulds and sieves.

Now compare it with artificial snow.

Figure 5.5 Matilda exploring Gelli Baff.

What does continuous provision for science learning look like?

Continuous provision is an essential learning experience in its own right; it should not be something to practise or consolidate what the teacher has input. Well-planned continuous provision allows children the time, space and equipment to develop those important scientific attitudes and skills. Valued, uninterrupted play supports the budding scientist and facilitates the uninterrupted experiences necessary for scientific enquiry. For young children, science enquiries should be exploratory, encouraging children to observe, raise questions about their observations, group similar objects, interpret their observations and communicate their ideas to others. Children need plenty of time to explore and undertake hands-on enquiries with time to go up blind alleys. They may make a mess but more importantly they have the opportunity to 'mess about' and engage in explorations and investigations (Hawkins 2002). Three-year-olds working with Lego will learn largely through trial and error and imitation; 6-year-olds can identify problems in their construction and learn from discussion and explanations. Time for exploration is important not only to become familiar with materials and tools but for questions to arise and be pursued. These experiences help develop strategies, skills and knowledge which can be seen as emergent science. This highlights the need for early years practitioners to have a sound knowledge not only of how young children learn science but also the ideas they may already hold: 'Children's concepts stem from and are deeply rooted in daily experiences which are helpful and valuable in the child's daily life context' (Duit and Treagust 1995: 47). In order to support children's learning, adults need to develop their own understanding of the science concepts so they have the knowledge to provide experiences which enable children to make the connections between what they know and what they are learning.

Task 5.2

Take some time to observe children in the different areas of continuous provision and ask yourself:

- **What science ideas are children developing?**
- **What are possible lines of development (PLODs)?**
- **How might you enhance this provision?**

Using areas of continuous provision to enhance early years science

Continuous provision permits children to explore any given scientific phenomenon across different areas, providing breadth and depth of experience from which they can generalize and learn scientific vocabulary, as discussed in

Chapter 1. It allows children to explore a particular science phenomenon within parameters set by the adults. It is not about having everything out all the time. Thus, when exploring 'pouring' children may start by pouring sand and water but then other materials can be introduced to enhance this. In the investigation area, children can explore how quickly a bubble moves through the different liquids (see Figure 5.6). Other materials such as rice, glitter, cornflour gloop or

Figure 5.6 Watching a bubble travel through a bubble bath

flour can be introduced to extend children's experience of things that pour. Noticing the sounds that these make as they are poured into different containers can be further explored in the music area. Case study 5.2 describes how Callum worked in the construction area to explore water flowing down pipes and tubes.

In Case study 5.2, we see one child working as a budding scientist being given the time, space and equipment to explore his own question. Callum sustained this activity every day for three weeks. It is important to notice that Callum felt that he could ask for what *he* needed and that the teacher provided the equipment he asked for. Here is an example where the teacher used what she knew about the child rather than being led by learning objectives. This is a point made in Chapter 9, describing planning for *endless possibilities*. Throughout this sustained investigation Callum is learning about properties of water: that it flows down not up; that the height affects the rate of flow, as does the diameter of the pipe.

Case study 5.2

Making sense of a problem that began at home

Callum (4 years)

On the nursery wall there was a two-weekly timetable of suggested activities for the sand and water play. I was about to set the water tray with water wheels and buckets. Callum came and stood beside me; I offered a water wheel for him to place in the water tray.

'Can I have the guttering and drain pipes instead?' We quickly lifted the box with the drainpipes beside the water tray. Callum rolled up his sleeves, put on a plastic apron and began to select what he wanted to use. His face was set as he became engrossed in his play (Figure 5.7). Other children passed by and watched for a few seconds, but did not join him. No invitations were given; he worked silently and alone.

As we tidied equipment away an hour and a half later Callum quietly put the pipes in the box.

At the review I asked Callum if he had enjoyed his activity and could he tell the other children what he had been doing. It transpired that his father had been trying to set up a water butt in the garden and had struggled to get the pipes to feed in to the butt.

The following morning Callum was waiting by the water tray ready to begin work. I asked if he needed anything. 'Something that will make the water turn a corner' and 'I need Peter to come and hold the pipe up higher for me.' The two boys set seriously to work, Callum orchestrating the proceedings. Again I left the boys alone during the play but asked them to share with us what they had been doing and if they had found anything out. Callum poured forth all that he had discovered.

'When Peter held the pipe up high the water came out more faster and if we used a big bucket it was faster still. The bigger [wider diameter] pipes were better

than the little ones [narrow]. Next time we want to go outside and put pipes together and make a real down spout!'

Figure 5.7 Callum and the drainpipe.

Task 5.3

Continuing learning about sound

Now using sound as the focus, what resources would you select for these different areas of continuous provision to provide a rich range of experiences to learn key concepts?

- Sand area wet/dry
- Water area
- Investigation area
- Construction area
- Creative area
- Music area
- Malleable materials area

NB It would be unrealistic to expect to enhance every area for any given science topic.

Key points

The characteristics of high-quality engagement in early years science are:

- It builds on children's prior experiences, backgrounds, and early theories.
- It draws on children's curiosity and encourages children to pursue their own questions and develop their own ideas.
- It engages children in in-depth exploration of a topic over time in a carefully prepared environment.
- It encourages children to reflect on, represent and document their experiences and share and discuss their ideas with others.
- It is embedded in children's daily work and play and is integrated with other domains.
- It provides access to science experiences for all children.

Opportune moments

The adult establishes an enabling environment for continuous provision. Many rich opportunities for sciencing, however, occur on a daily basis during unplanned events in the classroom and on the playground. These opportunities invite spontaneous sciencing and can lead to the 'having of wonderful ideas'.

'Spontaneous sciencing occurs whenever a child (or an adult) sees something of interest and wonders about it' (Kilmer and Hofman 1995: 55). A constructivist teacher recognizes such moments and pauses to observe, reflect and explore with the children. The occasion may be icicles hanging from the roof, a bird building a nest or ants at the base of a tree. By stopping to observe and reflect, the teacher takes advantage of these opportune moments to encourage children to grow in appreciation and understanding of the world around them.

What is the role of the teacher/adult in enhancing science in continuous provision?

The adult's role is clearly laid out in the guidance and plainly sees the adult as a facilitator once having observed the child at play. They are required to provide an 'enabling environment', one that 'provides stimulating resources which are accessible and open-ended so they can be used, moved and combined in a variety of ways' and 'ensure[s] children have uninterrupted time to play and explore' (Early Education 2012: 6).

Children need to engage in authentic enquiries that enable them to answer their own questions, which are appropriate and relevant to them if they are to develop scientific skills and strategies and see them as powerful and worthwhile tools that will enhance their knowledge. The teacher's role is to 'closely observe, document, revisit, and interpret the work of the children' (Hall 2010). Indeed, adults need to demonstrate that they value these enquiries and allow children to have a vested interest in what they are doing. Continuous provision allows children to choose where and how they want to play and investigate. Adults must be aware and respectful of this and realize the influence they can have on children's play spaces. They need to acknowledge that they bring an 'adult agenda' and this may change the nature of what is taking place (Canning 2007).

Adult intervention, when appropriate, can be a vital element for moving children forward in their learning. Knowing when and how to interact with a child who has chosen their activity in continuous provision is an 'art' developed through careful observation and a deep understanding of how children learn and make sense of their world. For adult engagement in the play to enhance a greater understanding of a scientific concept an adult must be an adept observer of play and understand its complexities. When invited in to a child's play or creating an opportunity to join the play, the adult should realize that they may not find an 'outcome'. Their involvement can enhance greater comprehension for the child but it may also create more questions.

Providing the guidance that children need to develop as investigative, critical thinkers is a challenge. Selecting the moment to intervene and scaffold children's learning is delicate. Rinaldi highlights this:

> The challenge for the teacher is to be present without being intrusive, in order to best sustain cognitive and social dynamics while they are in progress. At times, the adult must foster productive conflict by challenging the responses of one or several children. At other times, the adult must step in to revive a situation where children are losing interest because the cognitive map that is being constructed is either beyond or beneath the child's present capabilities. The teacher always remains an attentive listener.
>
> (Rinaldi 1998: 118–19)

It is essential that they consider carefully what will be the consequence of stepping in and have the confidence not to intervene. A colleague of ours uses the maxim 'Be the guide on the side not the sage on the stage.'

Adults need to remember that they act as role models for the child. Those who have a love of learning, are enthusiastic, question and enjoy new experiences are strong positive role models, while adults who are uninterested or less than enthusiastic about the world around them, who show no curiosity in things, will send negative messages about science.

Key points

How can the adult support science in continuous provision?

- Show genuine interest.
- Respect children's decisions and choices.
- Invite children to elaborate.
- Recap.
- Clarify experiences.
- Wait for a response.
- Observe children carefully before you intervene.
- Offer your own experience.
- Speculate.
- Use encouragement to further thinking.
- Model and demonstrate thinking.
- Remind.
- Reflect.
- Look closely.
- Find out.
- Identify.
- Add materials/equipment.
- Remove materials/equipment.

Meaningful conversations that support sustained shared thinking

In continuous provision, children choose and are engaged in investigations that are sustained by their interests and absorb them in ongoing open-ended activities. This provides the adult with opportunities to become involved in the children's thought processes by introducing lines of questioning and enquiry. Kath Orlandi in Chapter 2 talks about how children sharing their thoughts with other adults fosters growth in knowledge and understanding.

The role of the adult as a collaborator has been highlighted in *The Effective Provision in Preschool Education* (EPPE), affirming that positive outcomes 'are closely associated with adult–child interactions . . . that involve some element of sustained shared thinking' (Sylva et al. 2004: 6). Sustained shared thinking occurs when adults nurture and develop children's critical thinking skills. The adult begins with the child's interests and encourages a deeper thought process in order to 'solve a problem, clarify a concept, evaluate an activity, extend a narrative etc. Both parties must contribute to the thinking and it must develop and extend understanding' (Sylva et al. 2004: 6).

Sustained shared thinking can live and develop through children's scientific play. The adult establishes an enabling environment in the continuous provision that supports children's effective learning in science. This environment challenges the child's thinking skills and with the help and support of skilled adults they are given the opportunity to dwell in a deeper thought process. 'All of us have the ability to think creatively, but the extent to which we do will probably be highly dependent on the quality of our earliest experiences . . . with the right stimulus and support all children can learn to think in ways that enable them to solve problems, be inventive and make discoveries' (Bayley and Broadbent 2002: 1). Sustained shared thinking occurs most often when the child or children have initiated and led an activity. This is about building children's thinking and will, if supported by the adult, lead to 'the development of higher order structures of the mind' (Siraj-Blatchford 2009: 4). Rinaldi (2006: 126) describes this as a 'knowledge building process' which is best undertaken with others collaboratively and is concerned with children constantly rebuilding their understandings in the light of new experiences and ideas. When children do this thinking, learning becomes more involved and at a deeper level. It directly involves meta-cognitive processes. Adults can support this in practice by following children's interests, giving them time to develop and build on their ideas with few interruptions and seeing the mistakes children make as learning opportunities.

How do the questions you ask help children with their sustained shared thinking?

The questioning skills of the adult are crucial in scaffolding a child's grasp and understanding of scientific concepts during their play. Using questioning as part of an ongoing dialogue to permit the child to make sense of what is happening helps to create a genuine climate of enquiry. Loughran (2010) points out that, by reflecting on what works and why, teachers refine skills such as questioning. 'Learning to ask questions is an essential element of a child's learning. Questioning helps us to speculate in unfamiliar situations and solve problems' (de Bóo 1999: 14).

Task 5.4

Using questioning as a tool for promoting sustained shared thinking in continuous provision

Listen to questions asked by adults in the classroom. As you listen note the children's response in the right-hand column. Here are some questions to get you started. Add to them.

Table 5.1 What questions do adults ask?

Type of question	These can begin with:	How did the child/ children respond?
Closed questions	How many . . .? Is it (shiny/rough)?	
Open questions	Why do you think . . .? Why does . . .?	
Asking children to elaborate	Can you tell me more? Perhaps you can think of an example Why do you say that?	
Use positive questioning/statements	I am not sure; what do you think? That's an interesting idea I like what you have done there . . . what . . .? Have you seen what X has done? Why . . .? I never thought about that before You've really made me think	
Questions which encourage reflection	Do you think that everyone would think the same? I think . . . I wonder . . . . I imagine . . . . I agree I disagree	
Questions which encourage children to make links	When else have you seen this happening? Do all . . .?	
Questions to foster productive conflict by challenging the responses of one or several children	Can you see in the dark?	

Conclusion

Do not underestimate the power of continuous provision. It permits children to learn about the benefits, processes, attitudes and skills that are involved in scientific exploration and investigation. Children are more likely to become citizens who are ready, willing and able to join the scientifically literate community if they are treated as serious enquirers who want to find out about the world and are prepared to use all their abilities and resources to do so.

References

Bayley, R. and L. Broadbent (2002) *Let's Explore: 50 Exciting Starting Points for Science Activities (50 Exciting Things to Do)*. Birmingham: Lawrence Educational Publications.

Birth to Five Service (n.d.) *EYFS Definitions: Continuous and Enhanced Provision* (retrieved from www.birthtofive.org.uk/documents/Transition/Transition%20to%20yr%201/provision%20definitions%20_2_.pdf).

Bruce, T. (2005) *Early Childhood Education* (3rd ed.). Abingdon: Hodder Education.

Bryce-Clegg, A. (2013). *Continuous Provision in the Early Years*. London: Featherstone Education.

Canning, N. (2007) Children's Empowerment in Play. *European Early Childhood Education Research Journal* 15(2): 227–236.

Davies, D. (2011) *Teaching Science Creatively*. London: Routledge.

de Bóo, M. (1999) *Enquiring Children, Challenging Teaching*. Buckingham: Open University Press.

DfE (Department for Education) (2014) *The Early Years Foundation Stage*. London: DfE. ·

Duit, R. and D.F. Treagust (1995) Students' Conceptions and Constructivist Teaching Approaches, in B.J. Fraser and H.J. Walberg (eds) *Improving Science Education: International Perspectives*. Chicago: University of Chicago Press: 46–49.

Early Education (2012) *Development Matters in the Early Years Foundation Stage (EYFS)*. London: DfE.

Feynman, R.P. (1997) *Surely You're Joking, Mr. Feynman!: Adventures of a Curious Character*. New York: W.W. Norton.

Froebel, F.W. (1887) *The Education of Man*. New York: D. Appleton.

Hall E. (2010) What Professional Development in Early Childhood Science will Meet the Requirements of Practicing Teachers? *Collected papers from SEED Conference* (retrieved from http://ecrp.uiuc.edu/beyond/seed/hall.html).

Havu-Nuutinen, S. (2005) Examining Young Children's Conceptual Change Process in Floating and Sinking from a Social Constructivist Perspective. *International Journal of Science Education* 27(3): 259–279.

Hawkins, D. (2002) *The Informed Vision: Essays on Learning and Human Nature*. New York: Algora Publishing.

Kilmer, S.J. and H. Hofman (1995) Transforming Science Curriculum, in S. Bredekamp and T. Rosegrant (eds) *Reaching Potentials: Transforming Early Childhood Curriculum and Assessment*, Vol. 2. Washington, DC: NAEYC: 43–63.

Loughran J. (2010) *What Expert Teachers Do: Enhancing Professional Knowledge for Classroom Practice*. London: Routledge.

Morgan, N. and Saxton, J. (1991) *Teaching, Questioning and Learning*, Vol. 11. London: Routledge.

Rinaldi, C. (1998) Projected Curriculum Constructed through Documentation: Progettazione, in C. Edwards, I. Gandini and G. Forman (eds) *The Hundred Languages of Children* (2nd ed.). Westport, CT: Ablex Publishing.

Rinaldi, C. (2006) *In Dialogue with Reggio Emilia: Listening, Researching and Learning*. London: Routledge.

Siraj-Blatchford, I. (2009) Conceptualising Progression in the Pedagogy of Play and Sustained Shared Thinking in Early Childhood Education: A Vygotskian Perspective. *Education and Child Psychology* 26(2) (June): 77–89.

Sylva, K., E. Melhuish, P. Sammons, I. Siraj-Blatchford and B. Taggart (2004) *The Effective Provision of Pre-School Education (EPPE) Project: Findings from Pre-School to End of Key Stage 1*. Nottingham: DfE and the Institute of Education.

6

Inspiring early years science through role play

Jessica Baines Holmes

Introduction

> it is difficult to say, in a few and simple words, just how powerful a
> medium pretend play is for making and sharing meaning
>
> (Smidt 2011: 65)

When we step back and observe children at play, we see them recreate their experiences, enter into role and use props to represent aspects of their play. This type of play is referred to variously as pretend play, fantasy play, role play, ludic play (Hutt et al. 1989) and specifically, when others are involved, socio-dramatic play (Kitson 2010). While there are arguably differences between the definitions used by various writers, they all describe the type of play practitioners seek to facilitate in the role-play area of settings. There is no set script; it is play which evolves, changes direction and draws on children's past experiences and interactions.

From the 'Wendy house' of the 1970s' classroom, to the more gender-neutral 'home corner', a domestic role-play area is today a fairly standard feature in most early years settings. In many nurseries and reception classes this important domestic area is additionally complemented with engaging and stimulating role-play areas of both real and imaginary settings, reflecting the wider contexts that children have experience of in their communities (Rogers and Evans 2008).

It might not, however, be immediately apparent to practitioners how the provision of role-play areas can support scientific learning. This chapter aims to explore how science-oriented play can be encouraged in the role-play area by developing skills needed to work scientifically, enhancing conceptual understanding, supporting language development and harnessing the natural curiosity young children have about the world around them. The role of the adult sits at the heart of developing science-oriented play; the adult is both a facilitator and an intervener in play and these key roles will be considered. Throughout the chapter, consideration will be given to some of the challenges to overcome and potential pitfalls to avoid when developing role-play areas and enhancing learning in science.

The chapter concludes with some practical suggestions for creating and resourcing role-play areas which encourage science-oriented play; the skills and scientific concepts that can be developed are outlined.

What does role play offer?

Through role play children attempt to make sense of the world around them; they draw on their observations, their experiences, on stories they have heard and programmes they have watched and they delve into their own imaginations. Rogers and Evans (2008) comment that through role play children can explore notions of reality and fantasy simultaneously. It is this uniquely human ability to enter into a shared pretence that appears to support the development of a wide range of skills and abilities that contribute to later learning. The potential of role play in supporting children's cognitive development is posited by Whitebread and Jameson (2010), who report on the impact of play-world experience on developing narrative skills in young children, while in the Whitebread review on the importance of play (2012) the benefits in children's deductive reasoning and social competence are highlighted. Rogers and Evans (2008: 37) summarize key aspects of role play, stating that it:

- encourages representational thinking
- helps children to develop perspective-taking skills (seeing another's point of view)
- displays language competence
- involves problem solving
- encourages turn taking and negotiation.

These are all skills that support the young scientist. By developing a facilitative environment for role play, enabling opportunities for children to interact with each other and providing resources to encourage exploration and investigation we create rich opportunities to support scientific-oriented play (Johnston 2010).

Case study 6.1

From a student teacher

I based a role-play corner on the book The Owl Who Was Afraid of the Dark by Jill Tomlinson, which we used as part of our science learning journey. The owl is afraid of flying and the dark so we taught the children all about light through the book. We made a role-play area (a night scene from the book which had the treetop in it and the children designed the owl's home). The children tested out different light sources to find out which one was the brightest to help Plop see at night-time. We also designed reflective capes for him to wear so that when he was flying low to the ground cars could see him! We had made firework pictures and we put those up in the role-play area. We also made blindfolds for him by testing out which material blocked the most light so he could sleep during the day!

It was a great learning journey; the children learnt loads and they used the role-play area a lot to aid their learning of science.

In the example in Case study 6.1, the children initiated their own exploration of torches to solve a problem, they used the skills of classifying and identifying to sort fabrics and in their pretend play they began to use scientific language (light source, bright, dim, reflective, transparent and opaque) in a meaningful context. Alongside this, they developed their conceptual understanding of night and day and of light and dark, while common alternative conceptions were challenged e.g. that the moon or shiny objects are light sources.

A role-play context provides a risk-free environment to rehearse experiences that children have observed and to try out new skills and vocabulary. Davies (2011) suggests that the low-risk nature of such play can give children the confidence to be more inventive than they would be in a more structured activity.

In the night scene role-play example, the children were able to move effortlessly between pretend play and what can be termed exploratory play. It has been suggested (Hutt cited in Smidt 2011) that there is a difference between pretend play and exploratory play (what Hutt terms 'epistemic play') and that it is during the latter that learning occurs. Certainly most practitioners will find it easy to recognize the many opportunities for scientific learning in exploratory play: for example, investigating what magnets can do, sorting a collection of fruit and vegetables, handling and using hand lenses. Through the case study we see that, in the role-play area, there is the potential for both types of play.

For most practitioners, it can be argued that more important than categorizing types of play is having an awareness of the opportunities for learning that exist within the role-play area. Howe and Davies comment that the role-play area has the potential to be one of the most scientific parts of an early years setting. They raise the important point that 'the first stage in enabling the sorts of play likely to lead towards the development of scientific learning is to recognise where in our settings science may be going on' (2005: 156)

> **Key points**
>
> **In the role-play area children can:**
>
> - develop skills of working scientifically (including using equipment, raising questions, predicting, problem solving)
> - acquire and use scientific vocabulary
> - enhance their conceptual understanding
> - develop a positive attitude to the world around them and to ways of working scientifically.

The role of the adult in promoting scientific learning

There is a clear need for adults to equip children with spaces, scenarios and props to support their play in real and imaginary worlds (Bruce 2001). Kitson (2010: 118) suggests that it is the role of the adult 'to provide a structure within which the children can interact . . . to challenge and set up problems to be solved, to encourage children to test out ideas'. According to Howe and Davies (2010: 157), creating the right conditions and providing resources are necessary 'to maximise the potential for "science-rich" epistemic play'. Just how much needs to be provided in terms of resources is an issue that is considered later in this chapter. However, the role of the adult goes beyond that of provision. Harlen (2000) has refuted the notion that simply providing resources and allowing children to play will lead to children developing their scientific learning. Adult intervention is crucial if children's learning in science is to be enhanced through pretend play. The following section of this chapter explores this and builds on the points raised by Bruce (2001: 76) that the most important ways to support play are:

- to structure the environment, so that it is conducive to the occurrence of play
- for an adult to take an interest and be part of the play in a background way
- for the adult to be sensitive and aware of how to help things along without taking over.

The adult as facilitator: 'what will the role-play area be this week?'

Role-play areas will be most effective where they are co-constructed with the children, where relevant contexts emerge from the children's ideas and evolve and change with the ebb and flow of their interests. Children need a meaningful role-play area that echoes their real-life experiences; we need to ensure we are not asking them to imagine the unimaginable; children cannot play what they do not know. Edgington (2004: 218) argues that effective role play 'depends on children having had some firsthand experience of the context they are playing in'.

When children have had a trip to the market or a visit from a local vet they have real-life experiences on which to base their play (Figures 6.1 and 6.2). The domestic

Figure 6.1 Visit to the garage.

Figure 6.2 Garage role-play area.

'home corner' is prevalent in early years settings for good reason: it provides a familiar space that young children feel comfortable in. Through the use of visits and visitors and children sharing their experiences and interests we can provide children with the knowledge base that enables their creativity (Davies 2011).

Task 6.1

Following a workshop, an early years teacher said she needed to include more science opportunities in her role play. How might a visit by any of the following help this teacher provide a real-life experience on which children in her setting can base their play?

- Car mechanic
- Gardener
- Hairdresser
- Doctor
- Dentist
- Plumber
- Optician.

- Do they use science knowledge in their job?
- Which of these act as a scientist?
- How might this influence the teacher as he or she co-constructs the role-play area?

I recall a particularly well-used role-play area which was developed after a child in the reception class had visited the Science Museum in London. The child returned with a postcard, a packet of space food from the museum shop and huge quantities of enthusiasm, which in turn stimulated the interests of a whole group of children in the class. It was decided that a 'space station' would be developed in the role-play area. By watching clips of space walks on the television the children developed a shared experience on which to base their play. Questions and debate of what to wear in space led to discussion about properties of materials, gravity and temperature.

The space station role-play area developed out of the interests of the children and they were involved in the planning and construction of the area. The need to involve children in the development of role-play areas is supported by Rogers and Evans (2008), who draw on a number of studies which suggest that too much emphasis on adult-selected themes may limit children's interactions with one another and limit their imaginative activity. Broadhead and English (2005) researched the development of open-ended role-play areas; these became known as 'whatever you want it to be places', and led to Moyles (2005: 72) concluding that 'open ended role play has the potential to liberate the child's voice.'

What to provide?

According to Davies (2011: 37), 'The provision of interesting resources in the role play area can make all the difference to the scientific and creative potential of ludic play.'

Additionally, Johnston (2010) notes that developing understanding and promoting exploration can be encouraged by providing collections of objects in the role-play area. For example, in a woodland role-play area, a selection of plastic minibeasts can provoke discussion of where different minibeasts would be found (e.g. in the leaf litter, under a log) and can lead children to look closely at the number of legs and other physical features they have (it is, however, worth purchasing superior-quality, anatomically accurate, examples). The use of magnifiers, binoculars and digital microscopes can support the development of observational skills in encouraging children to look more closely at objects. Placing such equipment in relevant contexts within the role-play area offers children the space to handle, familiarize themselves with and develop their use of the equipment in a risk-free manner. When children later come to use the equipment to support their close observations in more a structured or guided activity, they are already familiar with it and it is more likely to be an enhancement of than a distraction from the focused activity (Johnston 2011).

Educational suppliers offer exact replicas of many resources for a wide range of role-play areas. These certainly do not provide the most cost-effective way to resource role-play areas, neither do they necessarily engage children's interests to the degree that real resources do. Goldschmied (2004) writes of the virtues of natural, real-life objects and the limitations of plastic in particular and argues that natural objects enable a multi-sensory approach to investigation which plastic cannot. In order to effectively harness children's interests through the use of props, we need to endeavour to use the real thing where possible (Edgington 2004). In the role-play garden centre, for example, provide actual plants pots, soil and seeds or seedlings for the children to plant. In the farm shop, a selection of real fruit and vegetables can be included for children to handle, arrange and sort. The careful inclusion of scented plants including lavender, rosemary or lemon balm among the actual plants provided in the role-play garden centre will also encourage exploration and close observation using all of the senses (Johnston 2010).

It can be argued that the simpler the resource, the more versatile it becomes. In the role-play cafe, a handful of conkers and acorns can represent any food type or even money, whereas a plastic hamburger or sausage will only ever be just that. The joy of open-ended materials is that they allow the children's imagination to make them whatever they wish and so facilitate a level of open-ended play that cannot be achieved with less flexible resources. The benefits of children adapting resources to meet their needs goes further. Rogers and Evans (2008) suggest that giving children flexible resources means that they need to problem solve to adapt the resources to their needs. This requires collaboration and decision making, skills which are an important aspect of working scientifically. In their research it appeared that an over-use of realistic props did not require the same level of agreement among players that was demanded by more open-ended materials and so limited the opportunities for children to develop those skills of problem solving and working together.

Indoors or outdoors?

Ideally, settings will offer opportunities for children to role play both inside and outside and to move, in role, between the two in a fluid way. This section considers the specific benefits offered by the outdoor provision of role-play areas. Outdoors, role play takes on another dimension: the play can occur on a larger scale; the greater physical space provides the possibility to use real resources. In the role-play 'builders' yard' children used scaffolding planks, milk crates and wooden boxes to construct structures, developing their understanding of forces, gravity and mass.

An unwanted aspect of role play often cited by practitioners is the high levels of noise. Noisy play can be a particular issue in reception classes where there may be a greater occurrence of adult-led activities; being outside removes this constraint and allows children to explore fully. Further findings indicate that outdoor play encourages children to create play spaces for themselves and to exercise greater choice over materials, location and playmates (Rogers and Evans 2008). It has been suggested that outside girls appear to take on more active roles and boys appear less disruptive to those around them and may experience fewer instances of conflict with adults (Santer et al. 2006).

The outdoor environment provides an open-ended flexibility which is difficult to achieve indoors; Harding (2005) comments that that both domestic and fantasy worlds can be explored within the same space. An example of this was in the role-play garden centre of a nursery where children were playing at watering plants with a hose pipe. The hose pipe, which had been left on the ground, was adapted by another child, who decided it was a 'snake'; this led to a group of children becoming explorers on safari, searching for wild animals in the garden area. Later that session, additional resources including binoculars, cameras and notebooks were provided to support and extend this play.

Some settings are fortunate to have a shed or similar structure which can act as a permanent role-play area, allowing props to be stored in all weathers. The many opportunities a permanent outside shed can offer are explored by Bianchi and Feasey (2011). For those settings without a permanent structure, the provision of simple, open-ended resources (plastic crates, tarpaulins, planks of wood, sticks) allow children to assemble den-like structures quickly. Hollow blocks can provide flexibility in creating boundaries and can be used to construct walls, enclose a space or become an object. During a recent trip with a group of children to a woodland area the children spotted some den-like structures which had been assembled by earlier visitors to the woods. The children worked collaboratively, developing these structures; some collected branches to 'fill in the holes' while others looked for pieces of wood to serve as chairs; another child used a fallen branch to 'sweep' the floor inside. The children worked purposefully, problem solving, negotiating and developing their knowledge and understanding of structures and appropriate materials (one child was overheard commenting on the texture of the tree trunks: 'this one is too "barky" to sit on – we want a smoother one'). The children then seamlessly moved into pretend domestic play, eating, interacting with one another and inviting visitors into their home. Similar opportunities can be facilitated by identifying natural spaces under trees, willow structures, or by placing crates and hollow blocks near gaps in trees to define an area which can be constructed and developed by the children. Resources and props can then be placed nearby to support the chosen theme.

The adult as intervener

As previously discussed, while the careful provision of a rich facilitative environment with appropriate resources is essential, to fully develop opportunities for enhancing scientific learning practitioners need to think carefully about how and when they can interact and enter sensitively into play. Through interactions with the children we create opportunities to find out more about their interests and their everyday ideas of the world around them. Through joining in the play, practitioners can take advantage of opportunities for sustained shared thinking (Sylva et al. 2010), engaging in shared conversations, problem solving and critical thinking. Johnston (2011: 30) states that 'early scientific development involves children in developing a wide and increasingly scientific vocabulary.' The adult entering into role with the children provides an ideal opportunity for children to hear and acquire scientific vocabulary in context. However, it is important to note here a cautionary point made by Smidt (2011: 12), who reminds us that adult intervention can be both 'helpful and harmful'. If intervention is not done with sensitivity and with regard to what the child is interested in it can easily destroy the play. Consider this point in regards to Case study 6.2, taken from a visit to a preschool room in a nursery.

Case study 6.2

I observed a small group of boys arriving at their nursery one morning and looking around deciding what to do. In the book corner, a large alphabet mat had been laid out; picture cards had been placed nearby. (Presumably the adult's intention was for the children to match the cards to the corresponding letter on the mat.) The boys noticed this large mat and, with enthusiasm, two of them climbed underneath it and called out to their friend 'Look, it's bedtime – we're going to bed.' The third boy at this point picked up a book from the shelf: 'then I must read you a bed-time story', he told the others.

At this point one of the practitioners hurried over, pulled the mat from over them and tried to show them what they 'should' be doing with mat and letter cards. It is hardly surprising that at this point the boys were no longer interested in this resource and instead wandered off in search of something else to do.

The timing of interventions needs to be carefully considered to ensure they 'neither intrude upon nor frustrate or terminate the play' (Santer et al. 2006: xvi).

Careful observation of children at play will support practitioners in familiarizing themselves with the dynamics of the children's worlds and play themes (Wood 2013) and will support making that potentially difficult decision of when to intervene and when not to. It is often the case that observations lead the practitioner to decide not to join the play at that moment, but instead to make a note of additional resources and equipment to provide the following day or of concepts to address at another time. For example, when two children were observed playing in the water trough in the role-play 'beach', a child's comments were noted: 'Don't put the spade into the water; it's too big – it will sink.' The following day a wider selection of resources of varying buoyancy (including small pebbles, large plastic beach balls etc.) were placed in the beach area to challenge the commonly held alternative conception that big objects always sink while small objects float.

Modelling

Exploration and investigation are at the heart of early years science. Watching, listening and hearing an adult modelling the skills of enquiry will benefit a child's scientific learning, (Sharp et al. 2012). Case study 6.3 demonstrates how adults can intervene to extend and develop play and use the opportunities created to model equipment and language.

Case study 6.3

From a student teacher

We had made a quiet part of the outside area into an explorer's camp. We draped a piece of fabric between two trees and we put a few boxes there and a camp bed, and a rucksack filled with hand lenses and binoculars. The children were all pretending to look for wild animals and there was a lot of running around and being scared of lions. Later, I arrived in role as another explorer interested in discovering minibeasts. The children watched me using the hand lens and bug viewers to look closely and then joined in (Figure 6.3). They became really interested in the woodlice and worms that we found near the camp and it led to lots more investigating and finding out about minibeasts.

Figure 6.3 Minibeast explorers.

Siraj-Blatchford comments that 'adults are role models for children, and therefore they have the power to influence values, attitudes and behaviour' (cited in Santer et al. 2006: 64). As practitioners, we can model the awe and wonder of seeing a spider's web sparkling with the morning dew or the frost on leaves. It is by modelling attitudes, language and use of equipment alongside the children that we can provide children with the knowledge and experiences which they can replicate in their own play.

As previously mentioned, there is a need to recognize that adult intervention is most likely to succeed in developing scientific learning where the practitioner is sensitive to children's interests and agendas; the aim is to become a partner in play. To achieve this we need to resist the desire to chase specific learning objectives through a barrage of closed questions which risk spoiling the play (Smidt 2011). Children's learning cannot always be planned for; the opportunities for scientific learning will often be unexpected and incidental. What adults can do is create a suitable environment which maximizes the opportunities for

Table 6.1 Some practical suggestions for role play

Role-play theme	Conceptual understanding	Skill development	Resource ideas	
			Inside	Outside
Garage	Forces Movement Friction Sound Absorbency	Testing Predicting Problem solving	Large hollow blocks Wheels Steering wheel Log book Hand pump/foot pump	Real tyres Hub caps Wheeled vehicles Buckets of water and cloths/sponges for washing Empty oil cans Telephone Tools – spanners/screwdrivers
Garden centre	Plants – recognize features of living things Parts of plants Plant life cycles	Classifying and identifying; sorting the fruit and vegetables in the garden centre, arranging the seeds according to their own criteria Noticing similarities and differences: growing scented plants (mint/basil) will encourage observation	Packets of seeds, selection of fruit and vegetable (real and imitation) Labels for plants Plants growing Garden tools – trowels/fork Overalls Seed catalogues Plant pots	Tyres filled with soil growing potatoes Growing plants Builders' tray filled with compost for potting, trowels, selection of larger pots Rakes/hoes/spades Large trolley for moving plant pots Watering cans Paving stones Seed trays Wheelbarrow
Seaside	Buoyancy Density Forces (floating and sinking) Sand – flow/mixtures/evaporation	Explore properties of wet/dry sand Look at similarities and differences Select appropriate equipment Make predictions	Buckets Spades Fishing nets Shells	Tarpaulin covered in sand/pebbles Paddling pool filled with water A boat/inflatable dingy – oars (or just a wooden wheel and boxes/crates) Snorkel/flippers
Builders' yard	Forces Movement Materials – their properties Health and safety Waterproofing	Explore and select materials	Clipboards Hard hats Tool box Selection of tools Pieces of pipe Tape measure Copies of architects' drawings Nuts and bolts	Empty paint cans Traffic cones/bollards Crates Guttering Planks of wood Bricks Umbrella/parasol

(Continued)

Table 6.1 Continued

Role-play theme	Conceptual understanding	Skill development	Resource ideas	
			Inside	Outside
Explorer's camp	Habitats Animals Camouflage Night and day	Using equipment Questioning	Periscope Binoculars Hand lenses Rucksacks Compass Fabric for sleeping bags etc. Pots and pans Lantern	Sticks and branches to build a camp fire Cooking pots Netting/fabric to camouflage camp Tunnels Boxes/crates for chairs/tables Torches Walkie-talkies
Dark woods/cave	Dark/light Night/day Light sources Shadows – investigate Nocturnal animals/habitats Animal adaptation	Raising questions Observation	Leaves Replica minibeasts Dark pieces of fabric	Bark chipping/leaves
Space station	Solar system Gravity Properties of materials	Raising questions Using equipment Comparing similarities and differences (materials)	Tunnels Shiny fabric/aluminium foil/black fabric/sugar paper Pictures of planets and stars Computer/if a tablet computer is available a solar system app can be used to navigate space and zoom in and out of planets Boxes	Rope (for space walks) Moon buggy
Opticians	Light Colour Ourselves Senses	Using equipment Testing ideas	Empty frames Eye chart Mirrors Hand lenses Concave/convex lenses Telephone Appointment book Coloured acetate	
Baby clinic	Ourselves Growth Health and hygiene Human needs	Measuring	Baby bath Scales Changing mat Charts Dolls Baby bottles	

science-oriented play and the first step towards this is practitioners being aware of where the potential for science is in the role-play area and how these opportunities might be maximized through modelling and sensitive intervention.

Conclusion

Johnston (2010) argues that exploration is most likely to be successful when it is purposeful and stimulated by artefacts, stories and play. Through the adult observing and listening to the voice of the child, role-play areas can be developed which harness the children's curiosity and interests and encourage children to explore. As Santer et al. (2006) comment, carefully timed adult interventions can support and extend play. Such interactions provide a wealth of opportunity to elicit alternative conceptions, to explore scientific ideas, to extend vocabulary, to develop skills of exploration and to promote a positive attitude.

References

Bianchi, L. and R. Feasey (2011) *Science Beyond the Classroom Boundaries for 3–7 Year Olds*. Maidenhead: Open University Press.

Broadhead, P. and C. English (2005) Open–Ended Role Play: Supporting Creativity and Developing Identity, in J. Moyles (ed.) *The Excellence of Play* (2nd ed.). Maidenhead: Open University Press.

Bruce, T. (2001) *Learning Through Play: Babies, Toddlers and the Foundation Years*. London: Hodder Arnold.

Davies, D. (2011) *Teaching Science Creatively*. London: Routledge.

Edgington, M. (2004) *The Foundation Stage Teacher in Action: Teaching 3, 4 and 5 Year Olds*. London: Paul Chapman.

Goldschmied, E. (2004) *People Under Three: Young People in Day Care* (2nd ed.). London: Routledge.

Harding, S. (2005) Outside Play and the Pedagogical Garden, in J. Moyles (ed.) *The Excellence of Play* (2nd ed.). Maidenhead: Open University Press.

Harlen, W. (2000) *The Teaching of Science in Primary Schools* (3rd ed.). London: Paul Chapman.

Howe, A. and D. Davies (2005) Science and Play, in J. Moyles (ed.) *The Excellence of Play* (2nd ed.). Maidenhead: Open University Press.

Howe, A. and D. Davies (2010) Science and Play, in J. Moyles (ed.) *The Excellence of Play* (3rd ed.). Maidenhead: Open University Press.

Hutt, S.J., S. Tyler, C. Hutt and H. Christopherson (1989) *Play, Exploration and Learning: A Natural History of the Preschool*. London and New York: Routledge.

Johnston, J. (2010) Exploration and Investigation, in L. Cooper, J. Johnston, E. Rotchell and R. Woolley, *Knowledge and Understanding of the World*. London and New York: Continuum.

Johnston, J. (2011) Learning in the Early Years, in W. Harlen (ed.) *ASE Guide to Primary Science Education* (4th ed.). Hatfield: ASE.

Kitson, N. (2010) Fantasy Play and the Case for Adult Intervention, in J. Moyles (ed.) *The Excellence of Play* (3rd ed.). Maidenhead: Open University Press.

Moyles, J. (2005) *The Excellence of Play* (2nd ed.). Maidenhead: Open University Press.

Rogers, S. and J. Evans (2008) *Inside Role Play in Early Childhood Education*. London and New York: Routledge.

Santer, J., C. Griffiths and D. Goodal (2006) *Free Play in Early Childhood: A Literature Review*. London: Play England.

Sharp J., G. Peacock, R. Johnsey, S. Simon, R. Smith, A. Cross and D. Harris (2012) *Primary Science: Teaching Theory and Practice*. Exeter: Learning Matters.

Smidt, S. (2011) *Playing to Learn: The Role of Play in the Early Years*. London and New York: Routledge.

Sylva, K., E. Melhuish, P. Sammons, I. Siraj-Blatchford and B. Taggart (eds) (2010) *Early Childhood Matters: Evidence from the Effective Pre-School and Primary Education Project*. Abingdon: Routledge.

Whitebread, D. (2012) *The Importance of Play: A Report on the Value of Children's Play with a Series of Policy Recommendations*. Brussels: Toy Industries of Europe.

Whitebread, D. and H. Jameson (2010) Play Beyond the Foundation Stage: Story-Telling, Creative Writing and Self-Regulation in Able 6–7 Year Olds, in J. Moyles (ed.) *The Excellence of Play* (3rd ed.). Maidenhead: Open University Press.

Wood E. (2013) *Play, Learning and the Early Childhood Curriculum* (3rd ed.). London: Sage.

7

Exploring toys and other resources to inspire science in the early years
Kathy Schofield and Lois Kelly

Introduction

Young children are endlessly interested in what they can hear, see and feel, so they are naturally drawn to explore how materials and objects behave and to find out what they do. The importance of providing children with the materials and equipment to 'mess about with' as they explore any given science ideas is highlighted by Hall (2010) and Harlen (2006: 10). An enabling environment for science plays a key role in supporting and extending children's development and learning so that they experience the pleasure of finding things out, which is fundamentally important to the human state.

In this chapter, we extend the discussion begun in other chapters about resources to inspire science in the early years. In Chapter 3, Babs Anderson makes some suggestions for books that inspire science learning. Faith Fletcher in Chapter 5 considers the opportunities that arise in continuous provision to enhance children's understanding of science and discusses how gradually introducing resources enhances that provision and extends children's thinking. In Chapter 6, Jessica Baines Holmes considers the props used to develop role-play areas with a science theme and Eleanor Hoskins looks at ways that technology can aid children's study in science in Chapter 8.

This chapter will explore the ways in which toys and 'other' resources inspire children's exploration of science ideas, influencing children's capacity to learn new skills, take an interest in and make sense of the world around them.

The purpose of resources

At the risk of stating the obvious, the purpose of toys and other resources is to provide children with the objects, materials, equipment and sources of information (Harlen 2006: 201) that stimulate their imagination and develop their scientific thinking. Worth (2010) points out that the resources determine the science phenomenon that children confront and manipulate, stressing the importance of ensuring that these are selected to enable children to focus on important aspects of a phenomenon. However, we need to be aware that resources that have been designed for a particular purpose can limit children's thinking (Davies 2011). As Jessica Baines Holmes points out, a plastic sausage can only be a sausage whereas acorns or conkers can be used to represent almost any food in a cafe role-play area. Similarly, if children use a manufactured marble run they do not have to solve the problem of how steep to make the slope or how to make the marble roll round corners (Worth 2010). Contrariwise, there are occasions when specific pieces of equipment are needed to support children's investigations. For example, hand lenses and nature viewers aid children's study of both living things and materials. To explore the properties of liquids children need a range of equipment such as sieves, funnels, droppers, syringes, jugs, bottles and watering cans with which and through which to pour the liquids. Children need to play with magnets to gain experiences which eventually lead to developing their understanding of magnetism.

The purpose of toys and resources then is to maximize opportunities for children to develop their knowledge and understanding of science. However, the notion that if simply provided with a set of resources, including toys, within a stimulating environment, children will spontaneously 'discover' scientific principles has been discredited (Fleer 2009; Harlen 2000). Adults are a key resource when they engage sensitively and thoughtfully in children's play because the discussions they have extend children's thinking (Davies 2011). For example, an adult observing children playing with dinosaurs in the sand tray introduced children to the idea of fossils and how they were formed (Fleer 2009). To have long-term value, opportunities for exploration and child-initiated learning provided by these resources need to be part of a carefully planned sequence and recognize that adult intervention which capitalizes on children's interest maximizes children's learning.

Case study 7.1

Science rucksacks

A resource with a purpose

Katie, the science co-ordinator at Anfield Early Years and Infant School, wanted to extend the opportunities children had to explore and investigate their world independently during play times as well as during the more formal 'work' times. To ensure that the children had ready access to resources that would enable this she developed science rucksacks. Each rucksack contained a variety of equipment: magnets, nature (bug) viewers, hand lenses, mirrors, slinkies, simple stop watches. To nurture the children's curiosity she also put in questions that could be used as starting points for children's explorations and investigations (Figure 7.1).

Figure 7.1 What is in the science rucksack?

Before introducing the science rucksacks to the children, Katie held a staff meeting so that the early years practitioners became familiar with the equipment in the rucksacks and ways that it could be used by the children. She then introduced the science rucksacks during an assembly. The science rucksacks have proved to be very popular with the children, who enjoy telling the adults each day what they have discovered.

For more information about science rucksacks see Bianchi and Feasey (2011).

Using toys and other resources effectively

Various factors need to be considered when thinking about making effective use of resources to inspire children's learning. One is how to achieve a balance between making the resources readily available for the children, avoiding sensory overload by having too many resources out and maintaining children's interest, because over time children lose interest with any set of resources (Beeley 2012; Johnston 2005: 4). As both parents and teachers, we discovered the benefit of putting toys away for a while because when the toys were brought out again the children played with them with renewed interest. Another is to consider the impact the resources will have on what children learn. For example an evaluation of an early trial for the Key Stage 1 science SATs found that one group of children came to the conclusion that all blue things sank because the teacher had not realized that all the objects she had selected sank just happened to be blue! Selecting the toys and resources which support children's learning is part of effective planning and is informed both by science background knowledge and previous experience of the resources which have supported children's learning.

Getting the right balance between making resources available so that children can initiate their own explorations and investigations, and selecting the resources which help children to focus on a particular science phenomenon is tricky. On the one hand, limiting the resources available to those that draw attention to the phenomenon might appear to limit child-initiated activity; on the other hand, having too many resources can be confusing and might lead a child to continue to explore a familiar concept rather than engage with a new idea. For example, if the focus is 'air' but the resources made available in the water area focus on 'pouring' children will focus their attention on 'pouring'. But replacing the 'pouring' resources with empty bottles, sponges, a turkey baster and droppers, all of which make bubbles when the air is squeezed out of them underwater, will draw children's attention to 'air'. Whenever new toys or resources are introduced children need time to play with them, to explore how they can be used. In addition children need to be able to select the resources *they* need to investigate the science phenomenon as Callum did (see Chapter 5). Having boxes of resources which you know children are likely to need, ready to bring out to support children's exploration and investigation of a science phenomenon, is good practice (Brodie 2013: 86; Hall 2010). What is important is that children know where to go to to find the resources they want.

Another factor to consider is how familiar the toys and resources are to the children. In Chapter 9, Di Stead makes the point that one reason for planning is to make effective use of children's time. This also applies when thinking about resources: if children are not familiar with the some of the resources they may need to be shown how to use them. For example, playfully showing young children how to squeeze the bulb of a turkey baster or a pipette to make bubbles in the water is likely to be more effective than waiting for them to discover this independently.

Key points

- Resources, including toys, are central to children's science learning.
- Children need to be able to select the resources they need when investigating a science phenomenon and to be introduced to resources that support their study.

Toys for understanding the world

In early years foundation stage settings, toys are an integral part of any classroom environment. Johnston (2005: 2) points out that when children play with toys they are developing a wide range of science ideas, so once in a while it is worth taking time to consider how different toys enable children's learning about the world and provide experiences that develop their understanding of science.

Task 7.1

A toy or some other resource?

1   Take a walk around your early years setting and list the *toys* the children are playing with. What science ideas or ideas about science are the children investigating?

2   Walk around the setting again and this time list *everything* the children are playing with. What science ideas or ideas about science are the children investigating?

3   Compare the two lists. What criteria did you use to decide if the object the children were playing with was a toy?

If you have completed Task 7.1, we wonder how closely your criteria for deciding if the children were playing with a toy matches the dictionary definition of a toy as:

1   A plaything *esp* for a child; anything intended for, or thought of, for amusement or pleasure rather than practical use

2   A thing or matter of little or no value or importance; a trifle

3   An imitation of something else as a plaything e.g. a toy dog.

(*Chambers Dictionary* 2002)

Something to think about

- Is a spoon a toy?
- Is it a toy when a baby bangs a spoon on a cup, bowl, saucepan or the tray of their high chair as they explore making sounds?

In this latter instance, the spoon meets two of the definitions for a toy because at that moment it is a plaything and it is an imitation of a drumstick.

How does the imagination of older children influence whether an object is a toy or not? We are sure like us you can think of many occasions when a stick has represented a sword, a light sabre, a walking stick, a crutch or a magic wand.

One problem with the dictionary definition is that it suggests, to the uninitiated, that toys are not to be taken seriously, which undermines the importance of toys in the cognitive development of young children and the impact they can have on children's understanding of the world. To effectively use any resource (including toys) as a catalyst for meaningful exploration, investigation and discussion, we need to ensure that their selection is thought through and has a clear purpose, while at the same time facilitating children's exploration of their world.

One of the characteristics of effective learning mentioned in Chapter 1 was ensuring that children gain experience of science concepts in a range of different contexts. Toys can provide that variety of experiences to support their learning. Many playground toys such as sit and ride toys, trolleys, seesaws, slides, climbing frames and swings can be a catalyst for children to mess about with forces and to talk about cause and effect (Figure 7.2). When playing on sit and ride toys or when pushing trolleys or buggies children investigate and explore:

• How they move on different surfaces

Figure 7.2 Sit and ride toys can be a catalyst for children to mess about with forces.

- How to make them go faster, slower or stop
- The difference in the force needed to make them go up or down a slope
- The difference in the force needed when there are passengers and no passengers on the bike.

Finding out how objects move on different surfaces can also apply to some rocking toys. You may have observed pairs of children working out how hard they have to rock to make the rocking horse, or a seesaw, slide across a smooth floor or the playground. Talking with the children as they play on the toys enhances their understanding of science ideas by encouraging them to think about cause and effect, or to think about forces, and helps children to make the link between the different activities. For example, when children push toys on different surfaces and when they slide down a slide they are experiencing the effects of friction. How might you help the children playing with the slinky make the link between how the slinky behaves and what they feel when they slide down the slide (Figure 7.3)?

Playing with toys enables children not only to develop their knowledge and understanding of science concepts such as forces, sound, light, materials and magnetism but also to develop more generic science skills, such as noticing (or observing) and comparing similarities and differences, which leads to developing an understanding of classification. For example, children learn to identify, sort and classify in a variety of different contexts when they play with toy cars, Lego, toy animals such as farm or zoo animals or sets of minibeasts. Playing with the toys with children provides practitioners with an opportunity to extend children's thinking by having a conversation about how the live animal is different from the toy animal or how a toy car is different from a real car. In Chapter 6, Jessica Baines Holmes suggests using plastic minibeasts in a woodland role-play area to develop children's thinking about features of minibeasts and their habitats.

In Chapter 3, Babs Anderson discussed the role of puppets to promote talk, but they can also help to reinforce children's learning in other ways. For example, having observed how snails move (Figure 7.4), children can then use a snail finger puppet to re-enact the movement, which reinforces their learning (Figure 7.5).

Messing about in science with balls

Young children love balls so let them investigate. Get all the different types of ball out of the sports cupboard and let them explore! How easy this sounds – it certainly uses children's interest as a starting point for learning science (Katz 2010) – but let's take a moment to consider what areas of science might be the focus of learning. Messing about with balls provides opportunities for child-initiated science enquiry and provide early experience of a variety of science concepts. In relation to science enquiry, they can:

- ask and answer their own science questions

Figure 7.3 The slinky works best on the stairs.

Figure 7.4 Observing a snail closely.

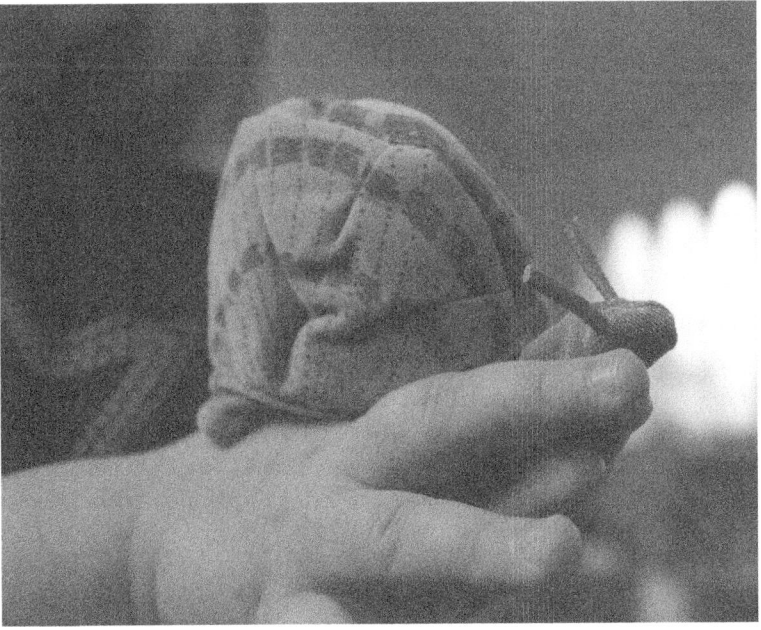

Figure 7.5 Using a snail finger puppet to mimic the snail.

- use their senses to sort balls into groups, looking for similarities and differences
- gain experience of different types of science enquiry (Turner et al. 2011).

In relation to science concepts, messing about with balls provides opportunities to:

- challenge children's naïve ideas about floating and sinking
- explore the effects of both gravity and friction
- explore how balls bounce.

As the children 'mess about' with the balls adults can observe what questions they are trying to answer and encourage them to make these questions explicit. A child may become engrossed in observing how different balls bounce, but not voice the question. Asking children what they would like to know about the balls can lead to a wide range of questions, some of which it may not be appropriate to answer and others which they can investigate. Children can find the answer to the question 'What's inside the ball?' if you are prepared, as one teacher was, to help children cut some balls in half. Another early years professional put a juggling ball in the selection of balls which she gave the children. This fascinated a group of children who, having felt the different balls, wanted to unpick the juggling ball to find out what was inside because it felt so different. However, questions such as 'why does a golf ball have dimples?' or 'why do balls bounce?' require complex answers about aerodynamics, the forces acting on the ball and energy transformations at the point at which the ball hits the surface, which are more suited to Key Stages 4 and 5. A suitable response to the question about why do balls bounce could be to look at how bouncy the different balls are.

As they play with the balls, children's attention can be drawn to the differences between balls such as size, colour and shape. They can be encouraged to sort the balls using these or other characteristics (e.g. squashy/hard, smooth/rough), introducing them to a sorting and classifying enquiry (Turner et al. 2011). As you and the children talk about the balls you could ask about the balls they play with in the paddling pool, with the intention that children play with the balls in the water tray, sorting them into those that float and those that sink. This provides an opportunity to challenge the idea, commonly held by many young children, that a large object sinks and a small object floats, because a marble will sink whereas a football or a beach ball will float. It is interesting to listen to associate teachers discussing whether an airflow ball is a 'floater' or a 'sinker'.

Providing a tray of sand can encourage children to examine the craters made when the different balls are dropped into it. They could find out how the height the balls are dropped from affects the size and shape of the crater.

Balls roll and balls bounce, and both of these characteristics can lead to further investigations. The question 'I wonder which of these balls is the bounciest?' provides the chance for children to make a prediction, to talk about what observations to make, how to record these and making it a fair test. Children will easily recognize that dropping the balls from the same height is 'fair' and that counting the number of times each ball bounces is easier than trying to measure the height of the bounce. A simple method for recording this is to use a different coloured sticker for each ball and to stick a sticker on a pre-drawn table for each bounce (see Figure 7.6). Counting the stickers, children review their predictions and answer their question.

Although the main focus for this investigation, with young children, is likely to be developing their ability to work scientifically, it is helpful to be aware that it provides early experiences about energy and forces, which help later learning.

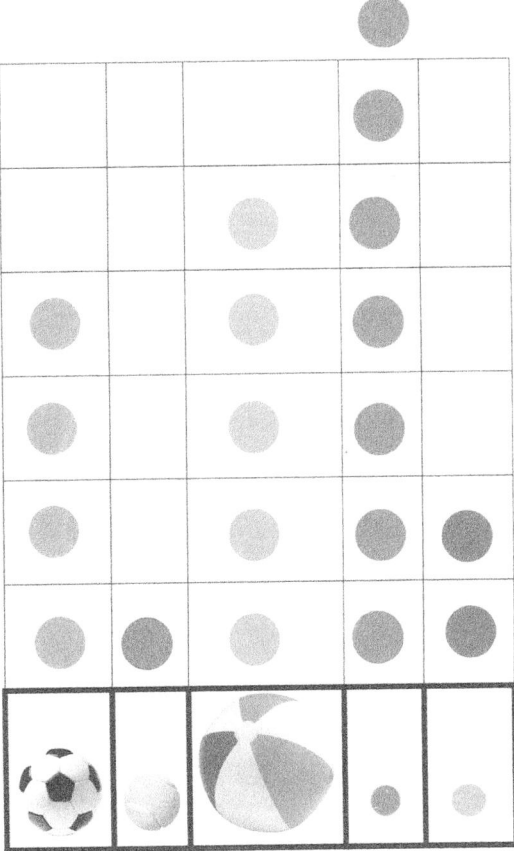

Figure 7.6 How many times does the ball bounce?

If children show an interest in balls rolling, then give them a challenge of rolling the balls down a ramp to knock over a skittle or ring a bell. As children investigate the effect of changing the height of the ramp and/or the size of the ball you will be extending their experience of cause and effect and encouraging them to talk about why things happen. A further line of development is to take this learning outside the classroom so the children can explore how balls roll on grass and the different playground surfaces. Children's thinking about the effect of slopes and inclines, friction and how different surfaces affect movement can be extended by introducing toy cars, which can be linked back to their experiences playing on sit and ride toys.

Messing about with mud

Another resource which is becoming a familiar feature in early years foundation stage settings is a mud kitchen (Figure 7.7), but if it is to be more than simply a play area and inspire science in the early years we need to have in mind the scientific experiences that engage and interest the children. An early years teacher in Wales, working with children aged 4–7, said:

> I feel the 'Mud Kitchen' has enabled the children to experience science at a different level and through a different medium. All children, regardless of

Figure 7.7 Mud kitchen.

their age and ability were able to access the materials and contribute to the ongoing discussions and investigations that happened naturally throughout.

Children in the mud kitchen will use their imagination to create a range of concoctions such as mud pies, magic potions, perfumes or soups. In the process, they gain experience of changes to materials caused by mixing and grinding and also separating materials. Found natural objects such as leaves, sticks, moss, small pebbles and stones provide children with the springboard to make magical cakes, stone soups or mud pies, while everyday kitchen equipment such as old saucepans, bun tins, bowls and a range of utensils such as old sieves, strainers, spoons, tongs, chopsticks and a pestle and mortar are some of the tools that support their explorations and investigations.

As previously mentioned, adults working collaboratively and engaging in discussions while the children are messing about in the mud kitchen can support the development of scientific ideas. For example, adults can initiate a conversation about the consistency of the mud pies or encourage children to notice what happens when more water or more mud is added. Wondering what will happen to the mud pies when they are left to dry will encourage children to revisit the activity and this experience of evaporation can be compared with drying washing, for example, which was discussed by Faith Fletcher in Chapter 5. Talking about what happens to the mixture when different ingredients such as sand or bark chippings are added can extend the activity. Children can also be challenged to find ways to separate the different ingredients in their mixture, for example by sieving. When thinking about the science experiences that the mud kitchen has to offer we need to consider how such experiences are different from, or extend, other similar experiences such as playing in the sandpit or with cornflour gloop.

The children's experiences in the mud kitchen should not be restricted to mud mixtures. Children could be challenged to make dyes by mixing leaves, such as dandelion leaves, with water and then pouring the mixture through a sieve. They may also try making perfumes.

Messing about with magnets

Magnets fascinate young children and are one of the early pieces of science specific equipment that young children are introduced to. They provide children with memorable experiences which are the foundation for later learning. So what might young children learn when messing about with magnets? They learn that only some materials are attracted to a magnet and, if the resources have been carefully selected, that not all metals are attracted to a magnet. Messing about with a wooden train set, Lois Kelly's sons experienced the repulsion of like poles of a magnet when they discovered they could make the carriages push away from each other as well as attract each other when making a train. Messing about with magnets, some children discover they can make some objects move as if by magic, for example by moving a magnet under a book or a table to make a coin move. In scientific terms, the children are experiencing the concept that the force of magnetism will act through non-magnetic materials. This can be adapted for the classroom by making a maze which is fixed to a clipboard or piece of corriflute. The children then make a 'bug' from a pipe cleaner, and using a magnet underneath the board they make the 'bug' move through the maze. A variation of this is to put some pieces of pipe cleaner inside a clean, empty plastic bottle, close the lid and make the pipe cleaners dance using a magnet on the outside (Figure 7.8). In Task 7.2, children experience that magnetism is a non-contact force and acts at a distance.

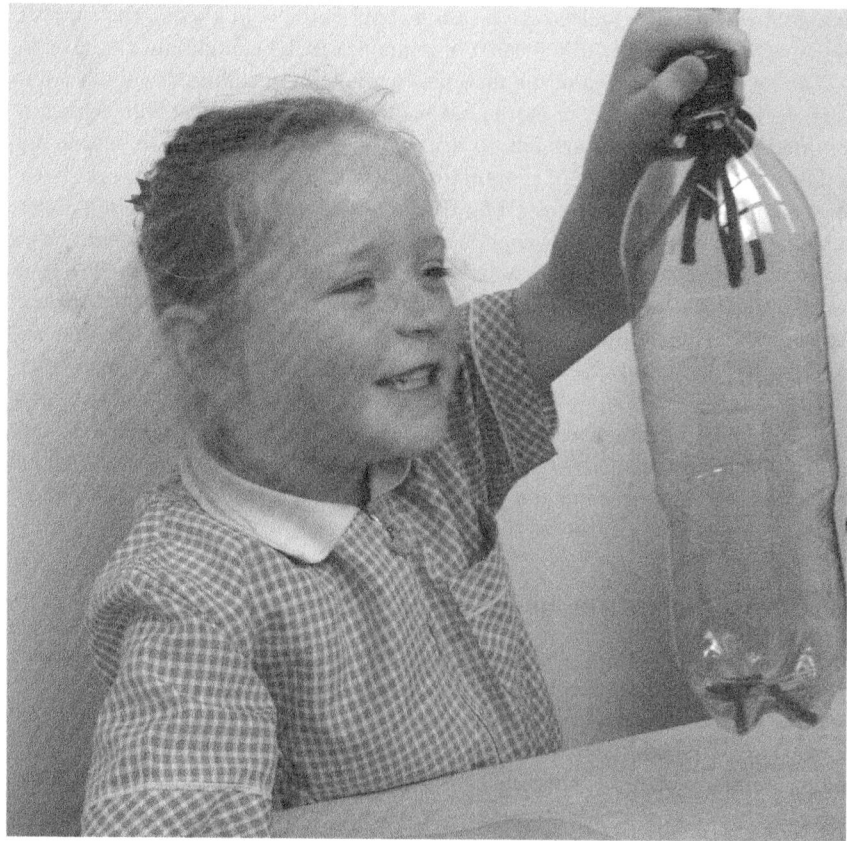

Figure 7.8 Look at the pipe cleaners!

Task 7.2

Charm a snake or fly a kite

You will need:
- Paper to make spiral snake or kite
- Thread (long enough to pull up with the snake/kite)
- Paper clips
- Magnets
- Small amount of Blu-tack.

Method:
1 Make either a spiral snake or kite with a tail out of paper.
2 Attach one end of a length of thread to the tail of the snake or kite and then using a small blob of Blu-tack attach the other end to a tabletop.
3 Put a metal paper clip on the head of snake or the top of the kite.
4 Hold a magnet near the paper clip on the snake or kite. Gradually lift the magnet higher; the snake or kite should rise up with it.
5 Discuss with the children why the magnet made the snake or kite rise.

Conclusion

In the final analysis, whether children play with toys, everyday objects, pieces of science-specific equipment, indoors or outdoors to engage with scientific concepts and ideas there will be no progression or development of understanding unless it is a collaborative process with a significant adult. The skilful intervention of early years practitioners in children's play is essential if children's thinking is to be challenged. This intervention will be informed by secure knowledge of the science ideas which are the focus of the activity and also knowledge of the children within the setting.

References

Beeley, K. (2012) *Science in the Early Years: Understanding the World through Play-Based Learning*. London: Featherstone Education.

Bianchi, L. and R. Feasey (2011) Science Beyond Classroom Boundaries for 3–7 Year Olds. Maidenhead: Open University Press.

Brodie, K. (2013) *Observation, Assessment and Planning in the Early Years*. Maidenhead: Open University Press.

Chambers Dictionary (2002) Edinburgh: Chambers Harrap.

Daves, D. (2011) *Teaching Science Creatively*. London: Routledge.

Fleer, M. (2009) Supporting Scientific Conceptual Consciousness or Learning in 'a Roundabout Way' in Play-Based Contexts. *International Journal of Science Education* 31(8): 1069–1089.

Hall, E. (2010) What Professional Development in Early Childhood Science Will Meet the Requirements of Practicing Teachers? *Collected papers from SEED Conference* (retrieved from http://ecrp.uiuc.edu/beyond/seed/hall.html).

Harlen, W. (2000) *The Teaching of Science in Primary Schools* (3rd ed.). London: Paul Chapman.

Harlen, W. (2006) *Teaching, Learning and Assessing Science 5–12* (4th ed.). London: Sage.

Johnston, J. (2005) *Early Explorations in Science* (2nd ed.). Maidenhead: Open University Press.

Katz, L.G. (2010) STEM in the Early Years. *Collected Papers from the SEED Conference* (retrieved from http://ecrp.uiuc.edu/beyond/seed/katz.html).

Turner, J., B. Keogh, S. Naylor and L. Lawrence (2011) *It's Not Fair or Is It?: A Guide to Developing Children's Ideas Through Science Enquiry*. Sandbach: Millgate House Education.

Worth, K. (2010) Science in Early Childhood Classrooms. *Collected Papers from SEED Conference* (retrieved from http://ecrp.uiuc.edu/beyond/seed/worth.html).

8

Using technology to inspire science in early years
Eleanor Hoskins

Introduction

Technology surrounds modern, young children throughout their everyday lives, and use in the early years has great potential to enhance educational opportunities (Siraj-Blatchford and Whitebread 2003). Nutbrown (cited in O'Hara 2004) and Hayes (2006) advocate that the only way to prepare children for tomorrow is to give them what they need today, allowing them to make informed choices with technology. Because of the immersion that everyday life brings, young children do not find the pace and change of technology as confusing as adults do (Johnston 2005), which makes the exposure to technology at a young age very important.

Young children are constantly absorbing information from their everyday life; it is an interpretation of *their world*. The use of technology offers the opportunity to expand this world, which provides great potential for young children to explore science in the world around them (Qualter 2011). Technology can allow young children to experience phenomena which otherwise would be unavailable or impossible, such as observing plant growth or a changing butterfly through time-lapse photography (Johnston 2005), thus opening new doors to exploration and investigative learning which would normally be out of bounds. The ability to experience the unavailable and impossible is a major influencing factor when considering the use of technology to enhance early scientific exploration. As Howard et al. (2012) highlight, the good use of technology can enable access to resources that would not otherwise be viable (because of cost, time or safety) in the classroom. Such restrictions can limit what opportunities are available to young children and technology can help overcome these limitations and restrictions.

Technological innovation and expectations with young children

Any doubts about young children's ability to access technology need careful examination. Educators can often underestimate young children's experience and confidence with technology (Kennington and Meaton 2009) and their abilities are often misunderstood (Johnston 2011). Miller (1997) proposes that the main barrier preventing children from having the opportunity to make decisions is the widespread belief that they are not capable of making informed decisions and that adults know best.

Low expectations about young children's capabilities need to be challenged. It is important to remember that, by the age of three, children can generally express themselves fairly clearly and are becoming adept at solving everyday problems (Lowe 2009). As a young brain develops, a vast number of connections between brain cells form in the first few years of life causing the human brain, uniquely, to quadruple in size between birth and six years of age (Whitebread 2008). The development of these connections, which is the very essence of learning and thinking, is reliant on stimulation (Katz 2010). Stimulation is an imperative factor considering the age-related connections. Typically, our optimum level of stimulation decreases as we get older. Therefore, the chances are that if an adult is feeling comfortable with an activity then some of the children will be bored (Whitebread 2008). This research suggests that young children should be provided with opportunities to experience a wide range of situations, to ensure optimum stimulation. In terms of technology, this equates to a range of equipment, opportunities and first-hand experiences which will enable ultimate access. Enabling ultimate access is imperative; Feasey and Still (2006) recognize that access to a wide range of applications provides potential for young children to become literate with technology from a very early age. Such access to a wide range of applications provides the essential stimulation that our young early years children need. The 'access' and 'opportunity' elements are pivotal since with young children being truly 'active' learners, they do not just learn what they are taught; rather, they learn what they experience (Whitebread et al. 2008).

In 2009, the Executive Director for Strategy and Communications at the British Educational Communications and Technology Agency (BECTA) visited Homerton Children's Centre in Cambridge. Upon arrival Mr Richardson found himself being photographed with a digital camera and, within a short time, this image had been loaded onto a computer and printed as a poster with a label. 'It is not unusual for schools to give their guests an enthusiastic welcome, but what's perhaps surprising about this story is that the children in question were only three and four years old' (Furness 2009: 1). We must never underestimate young children's abilities to absorb new information and to cope with new ideas (Whitebread 2008). Young children need exposure to a range of technologies at a young age to help with the stimulation and development of brain connections. This, in conjunction with high expectations and faith in young children's ability, could encourage and support children to push the boundaries of their technological understanding.

Using technology to enhance early science skills

Observation enhanced by technology

As Johnson (2005) recognizes, observation is arguably the most important skill in science and certainly the first skill that is developed. It is therefore a fundamental aspect of learning that is possibly taken for granted throughout everyday life. Within early years, the power of developing young children's observational skills cannot be underestimated. In terms of other scientific skills, observation can be seen as the initial building block because it provides a good starting point for further exploration and investigation. In addition, it develops the ability to use all senses appropriately, which, in turn, enables learners to find patterns while developing the ability to sort and classify (Roden 2011).

Close-up observations and captured images help young children to focus on the matter carefully, allowing them to slowly develop essential observational skills. An excellent technological tool for supporting this process with young children is the 'Easi-scope', which is a small, egg-shaped, hand-held digital microscope that can be connected to a laptop via USB.

The LED light within the microscope clearly illuminates the observed object and the one-click top button captures images with ease and minimal fuss. The easy and safe features of this durable technological device, which is also now available as a wireless product, lend themselves to young children and the encouragement of independent observation. Independent observation is imperative for exploration, since young children who are often very perceptive and creative observers are given the opportunity to observe freely without imposed beliefs, which in turn will help them develop confidence in their observations (Johnston 2005). In addition, due to the easy and safe use, the technical device also encourages young children not only to *observe* but to operate *independently* without direct adult help (Figure 8.1). Young children are perfectly capable of manoeuvring the device and capturing images and this in turn supports their development of prime personal and physical skills. Such features mean the microscopes can be set up within role-play or investigative areas for the children to use frequently to enhance their scientific exploration. An example of this is given by Linda Atherton in Chapter 4, where a digital microscope was a starting point for children's investigations into what was inside their ears, and for looking closely at a towel.

The 'Easi-scope' enhances early years science as it allows young children to observe detail within objects so clearly. Most importantly, the feature that allows a one-click snapshot ensures a clear image can be captured and observed in further detail without the worry of young, wobbly hands. Such captured

Figure 8.1 A hand-held digital microscope helps children look closely.

images can be used for further observational drawings or as a reference for future observations and explorations. In addition, the captured images can also be utilized to extend the observational process further by the use of an interactive whiteboard (IWB). Images captured by young children during explorations with garden snails, for example, can be displayed for all to see clearly using the IWB. Opportunities for all the children to collectively observe and identify basic features such as 'shell', 'shell coil', 'tentacles' and 'mouth' suddenly become available. Such use of the 'Easi-scope' allows technology to enhance the science through large-scale, detailed display. Through the projection of real-life images, young children can be supported to collectively identify, which in turn develops their specific scientific vocabulary.

Young children's observations should take account of their whole world. They should therefore involve using other senses as well as sight, provide opportunities for children to know about similarities and differences (DfE 2012: 9) and allow young children to identify sequences and events in the world (Johnson 2005). Therefore an excellent opportunity for young children to develop observational skills is to explore outdoors where they can see, smell, hear, touch and, where applicable, taste. Learning outside the classroom builds bridges between theory and reality (DfES 2006) and allows young children to extend experiences within *their world*. With early years science, these first-hand experiences not only provide ample opportunity to develop observational skills but also allow young children to develop their understanding of the diversity of living things, which in turn reveals the differences there are between things that are similar (Roden 2011).

Exploration of habitats and pond dipping with young children embraces the benefits of outdoor learning, as the children use their senses to *hear* the birds, *smell* the fresh air, *touch* the pond vegetation and *view* a surrounding environment in reality. While pond dipping with young children it is important to remember that their safety and care are paramount, but it is also imperative that they are provided with the opportunity to closely observe to the maximum. A waterproof camera such as the Snake Scope, which has a USB connection and flexible long neck, allows images to be transmitted from below the pond surface, thus permitting the otherwise unavailable to become available. The long, flexible neck of the water-proof camera allows supervised children to safely search under water and observe creatures within their natural habitat. Clear detail about underwater activity can then be accessed to supplement above-water observations, which in conjunction with registered sounds and smells helps to encourage young children to absorb the whole environment and observe thoroughly. The easy-click button captures images, much like the aforementioned digital microscope, that can be observed in further detail alongside or after the activity, which helps to elongate the learning experience for the children back within the setting. The technology of a flexible-necked, waterproof camera most certainly enhances outdoor scientific learning with young children since it opens up opportunities to make observations that would otherwise not be possible (Porter 2013). Without such a device children would certainly have limited opportunity to safely explore and observe under-water habitats as they present themselves naturally.

Observation of similarities, differences and classification

The core task of recognizing similarities and differences between objects and events and to rearrange them through the classification process, according to features they have in common (Johnston 2005), is very important with young children. One imperative reason for this is because it is generally recognized that learners find it more difficult to notice similarities because differences are more obvious (Roden 2011).

A Bird Cam uses technology that can record events over time, and the establishment of such technology with young children encourages an introduction to scientifically enquiring 'observation over time'. While observing over time, young children are provided with the opportunity to focus on similarities and differences in detail and this can relate to the development of classification skills. The constant focus of a Bird Cam, especially if it records motion-activated photos, provides an ideal opportunity for young children to constantly look at similarities over a short space of time alongside huge differences over a wider stretch. Such technology permits images to be captured and recorded, thus allowing young children to observe bird activity accurately and safely. In this instance, technology once again enhances core scientific learning by providing opportunity to recognize similarities and differences while overcoming the natural restrictions of young children 'physically' observing an inhabited bird box at close range.

The use of digital images is an excellent way to record the results of investigations and to illustrate practical activities (Porter 2013), which makes the use of digital cameras with young children very important. Young children can easily capture images of seed germination and growth or seasonal weather changes to note developments without any need for further recording. While young children were working on an 'enchanted tree' project, they visited the same tree frequently throughout the year. At each visit, the children captured images of the tree and its surroundings. This provided opportunity for close observation of the lush, green tree throughout summer, the changing leaves of autumn, the bare skeleton of winter and the fresh blossom of spring. Since young children do find it more difficult to notice similarities (Roden 2011) a study of the same subject, in this case a tree, that remained *similar* for long stretches of time helped to develop their attention to detail. The use of digital cameras enhanced this process as the children captured regular images to create a photo diary. This collection of images ensured that the children could observe regular similarities and then note small variations over time as the tree progressed through the seasons. Without the digital images, the children would not have been able to keep an accurate, visual record, which ultimately helped to develop their observation of small detail.

Taking the innovative leap: ensuring optimum use of technology to enhance early years science

Interactive whiteboards are a technology used by many schools and educational settings. Such display technologies are important and hold many benefits but there is scope for encouraging more engaging and interactive forms of teaching and learning with technology. One limitation of IWBs is that they do not cater for individual learning because their primary purpose is for teachers to use in whole-class situations (Howard et al. 2012) and over-use of the IWB could result in learners passively watching moving images while merely waiting for a turn to join in and touch (Ward 2011).

Ofsted (2009) stated in their *Importance of ICT* report that foundation stage children were very confident using the keyboard, mouse and onscreen menus. For young children such skills with technology are important, but it is also imperative that early years practitioners remember that technology must enhance learning and therefore it is not just any use of technology that is important but how it can be used effectively and with purpose (Howard et al. 2012).

In terms of using technology 'effectively and with purpose' within early years science it is vital to remember that with any science activity the science must be the priority (Harlen and Qualter, cited in Howard et al. 2012) and therefore the technology must enhance the science.

QR coding to enhance early years science

Many schools and settings are now using tablet computers with touch-sensitive screens. Such devices are easy for young children to access and are often familiar as a result of advertising and home experiences. Tablets can have application software (apps) loaded onto them which can be used to support learning. There are some good apps that can be used to support early years science, especially in terms of recognizing similarities and differences in relation to living things and places. The clear photos and distinct sounds from a range of apps can complement a selection of early years learning topics. Within early years science, there are several good apps featuring zoo, farm and marine animals that can successfully support an introduction or follow-up to an educational visit.

However, using tablets in conjunction with quick response (QR) codes can be a much more innovative way to use such devices with young children.

QR codes: essential information

What are they?

A QR code is a type of two-dimensional (or matrix) barcode. Like standard barcodes you might find in any shop, QR codes hold encoded information that is revealed when scanned by a special reader. QR codes can be used to store greater detail than a standard barcode and are typically used to directly link to website addresses, images and contact details.

How do you read QR codes with a tablet?

Once a QR code reader app such as *i-nigma* is downloaded onto a device like a tablet this can then be used in conjunction with QR codes. The tablet's camera can simply be pointed at a printed QR code for instant recognition and transportation to an intended website, text, email or image.

How can codes be generated to be used with children?

QR codes can be generated through various free websites (QRStuff 2013); alternatively, a QR code generator app can easily be downloaded. Information about a website, image or video can then be entered and a QR code can be generated and printed off. These can then be used by children with a device such as a tablet (with installed QR code reader app) to point at the printed code and link to an intended website, text, email or image immediately.

Focusing on *how* QR codes can enhance early years teaching and learning it is important to remember that giving young children immediate access to a pre-selected website or image has four clear advantages. These attributes make this approach to Internet learning:

- *Accessible*: young children can easily manipulate and point a tablet (with QR code reader installed) at a printed QR code.

- *Instantaneous*: the immediate response means young children are not wasting time typing in web addresses that can prove tricky or unsafe and can interfere with limited attention spans. The instant recognition takes the young children to an *intended* image or website immediately.

- *Easy*: the process of 'point and go' is so straightforward that young children are no longer reliant on adults typing in web information, which in turn promotes independence.

- *Safe*: young children are immediately transported to an intended website or image without any Internet roaming.

To focus on QR coding enhancing early years science, young children can easily use the codes to accompany a trail such as a minibeast hunt. Before the hunt, QR codes that link to key minibeast images or websites can be generated and printed out. These QR codes can then be displayed at several intervals throughout the hunt close to certain minibeast habitats. As the children discover creatures such as snails or woodlice throughout the hunt, they can scan the displayed QR code with a tablet and immediately retrieve accessible, first-hand information about the minibeast to accompany the discovery and enhance the learning activity.

QR codes allow the technology to enhance the learning and not distract from it. In terms of scientific exploration, using this instantaneous technology supports young children in seeking information while exploring, thus making the scientific and technological learning simultaneous (Figure 8.2). It therefore avoids any constant searching that can soon amount to a process where the child begins to flick through information, making little sense of anything they are looking at (Harlen and Qualter 2009) and keeps the subject-knowledge intention focused. This 'simultaneous' experience of technology and scientific exploration provides a paramount example of how technology can enhance early years science.

4D immersive space

4D immersive spaces, also known as creative classrooms, are created through LED lighting, giant projection and surround sound. These elements can be changed

Figure 8.2 Using QR codes to identify trees.

and manipulated to instantly transform and create a unique teaching and learning environment. 4D immersive technology is a remarkable resource that can replicate experiences such as the gradual change of seasons. Young children can instantly engage with the changing colours, shapes and sounds throughout the seasonal year within one activity. While engaging with seasonal 4D sights and sounds young children can supplement these experiences with other activities such as selecting appropriate clothes to wear, making appropriate drinks or choosing hats for the correct time of year. Closely linked to the seasonal experiences is a focus on the weather. Observing and experiencing the weather is an important aspect of early years science and often an unpredictable one. With the use of 4D space all weather can be guaranteed, which means even a snowstorm can be generated on demand through projected images and sounds. Again, within one focused activity, young children can experience the sights and sounds of sunshine, rain, snow, wind and hail, which helps to enhance their observation of similarities and differences as well as to provide a unique, inspiring experience otherwise impossible.

Many schools and settings are working to install immersive spaces and establishing this approach to teach science as well as other curriculum areas. Networks are also being set up between schools and settings with 4D immersive spaces to share ideas and best practice and to inspire others.

Use of Skype to enhance early years science

Skype is a Microsoft software application that allows users to communicate through voice and video calls using a device with a webcam via the Internet. In terms of enhancing early years science this technology can be used by a group, led by an adult, to keep constant communication throughout a learning topic.

Focusing on animals and plants, since they are obvious foci for early years science, links can be made with a zoo, farm or botanical setting to keep regular contact between the children and the outside world. Specifically, once a new animal is born at a zoo or farm the children, once arrangements have been made by an adult, can access regular, visual updates about the animal via Skype. This enables the children to ask questions, talk about changes and to also understand essential similarities and differences over time.

The use of Skype is also an essential tool following an actual educational visit. On a visit to a farm, for example, young children observe, use their senses and gather the first pieces of information. Following the visit, previous learning can easily be capitalized upon by keeping regular communication with the visited farm. This allows young children to process previous information then ask further questions and make new comparisons. As Hoath (2012) recognizes, serial visits to the same venue are likely to result in long-term benefits, although set against this is the time away. The use of Skype technology solves this problem since it provides young children with the opportunity to keep a thread of learning going without repeatedly leaving their educational setting to make a journey that is not always time or cost effective. However, this does not mean that Skype should ever replace outdoor learning or educational visits entirely. It is well known that quality learning experiences in 'real' situations have the capacity to raise achievement (DfES 2006: 3) and therefore technology such as Skype should only ever be used to enhance rather than completely replace the out-of-setting learning.

As a supportive tool, Skype contact can support educational visit experiences by keeping memories alive. The use of Skype enables the children to maintain a visual link and provides an opportunity to monitor changes long after the visit date. In preparation for Skype sessions, the children can talk about what they want to find out and decide group questions. This in turn enhances their science understanding through developing questioning skills while simultaneously encouraging prime communication opportunities.

Last of all, although it may seem an obvious point, the issue of safeguarding is paramount with any communication via the Internet. Skype activity should always be adult led within a setting and young children should be made aware of this fact. Opportunities to teach about safe Internet communication should be taken while working with Skype and this should run simultaneously alongside other learning. In addition, consent from parents and guardians should always be sought before embarking on a Skype project. Further information about Internet safety and how to present advice can be accessed via the document *Advice on Child Internet Safety* (UKCCIS 2012).

Role-play areas

Role-play areas are an obvious feature within the early years environment. They should replicate the world around us, and since technology and its use in the early years have great potential to enhance educational opportunities (Siraj-Blatchford and Whitebread 2003) an obvious and innovative response is to link role-play areas with technology. A *police investigation laboratory* encouraging young children to observe fingerprints and other clues using digital microscopes will generate guaranteed interest from the children. In addition, an *under the sea* or *rainforest* role-play area can be enhanced with the use of sound apps to replicate not only the sights but the sounds of these environments.

If a setting has sufficient heat and light then a replica *greenhouse nursery* can make a successful role-play area. While in this area young children can plant, grow, nurture and therefore observe. The use of digital cameras can enhance the experience as children independently collect images of the growing plants to monitor growth over time. In addition, children can be encouraged to collect images to contribute to a group or class brochure for 'plant sales'. Obvious, clear links with the outside world are imperative for early years role play and technology makes these links realistic.

Reflective summary

It is clear that the use of technology, when used as a pedagogical tool, has the potential to be powerful. However, when planning to use technology to enhance early years science it should always be considered *how* it will be used to its ultimate potential. The following questions can be asked by practitioners during the embryonic planning stage:

- Will it enhance the learning activity for the children?
- Will it help to engage children at a new level?
- Will it overcome restrictions related to safety, time or cost?

Most importantly, it should also be asked:

- Will the scientific learning intentions remain paramount?

Building on these reflective questions, the following points summarize key elements from this chapter to focus activity ideas and prompt further thinking:

- Ensure the use of technology still allows the early years science to remain paramount.
- Do not underestimate young children's ability to access and operate technology confidently if they are nurtured carefully.
- Encourage young children to use technology independently where it is possible: for example, within investigative or role-play areas.
- Be innovative with technology as a setting or school and try out new ideas to capture the children's interest as they explore.
- Make links with the outdoors and use technology to create situations that would not otherwise be possible.
- Use technology to create links with another location and build on learning stimulated by an educational visit.

Technology is a wonder that is forever evolving. Reflecting on my own training and experiences since I was a newly qualified teacher, it is remarkable how things have changed in just 15 years. The first classroom I worked in housed one computer in the corner that children took turns to access. There were no digital microscopes or cameras, laptops or Skype. This one computer was the only technology available to support and enhance any teaching and it was the same story in many classrooms around the country. The introduction of school 'computer suites' soon solved limited resource problems as they enabled all children to experience computing without waiting for turns. But suites were a giant step away from the essential 'simultaneous' technological and experiential scientific learning highlighted in this chapter. These days tablet computers, laptops, digital cameras, voice recorders and hand-held digital microscopes in schools and settings ensure abundant technology availability and widespread opportunities for simultaneous, mobile learning. In addition, the growing use of innovative technology such as 4D immersive spaces provides opportunities that were only visions and dreams a short while ago. In just 15 years, technology has progressed in giant leaps with technology, ensuring that children these days can have the best learning experiences possible. Take some time to contemplate these giant leaps in technology and the impact they have upon teaching and learning within early years science today.

Technology offers multiple opportunities and mobile experiences, and often overcomes safety and time restrictions related with science especially within early years. So now, consider how this can be utilized and capitalized upon within your school or setting.

Task 8.1

Putting ideas into practice

Why not try . . .

- Making links with a local animal setting or garden area to arrange regular Skype sessions for your group or class of children alongside an educational visit.
- Encouraging children to independently use tablets and QR codes within the investigative area: for example, set up immediate links to images or simple websites about seed planting.
- Using tablets and QR codes to accompany a minibeast hunt, materials search or plant discovery trail to support instant learning with Internet websites and images.
- Using apps to add photos or sounds to a science role-play or investigative area.
- Creating role-play areas that imitate a laboratory, under the sea world or greenhouse nursery in conjunction with digital microscopes, sound apps and digital cameras to enhance the play.
- Using a Snake Scope to observe underwater activity within a pond or rock pool, or in the sea.
- Setting up a bird nest box with a Bird Cam to monitor development and broadcast images via the school or setting's website.
- Using digital microscopes with the children to investigate inside unusual fruits such as dragon fruit, kumquat, passion fruit and pomegranate.
- Experimenting with time-lapse photography to observe change over time, such as a seed geminating, a butterfly emerging from a chrysalis, a chick hatching.
- Using a digital camera to keep a series of images to record seasonal change.

References

DfE (Department for Education) (2012) *Statutory Framework for the Early Years Foundation Stage: Setting the Standards for Learning, Development and Care for Children from Birth to Five*. London: DfE.

DfES (Department for Education and Skills) (2006) *Learning Outside the Classroom Manifesto*. Nottingham: DfES Publications.

Feasey, R. and M. Still (2006) Science and ICT, in M. Hayes and D. Whitebread (eds) *ICT in the Early Years*. Maidenhead: Open University Press.

Furness, V. (2009) *Curriculum – ICT – From ABC to ICT, TES Connect* (retrieved from www.tes.co.uk/article.aspx?storycode=6014565).

Harlen, W. and A. Qualter (2009) *The Teaching of Science in Primary Schools* (5th ed.). Abingdon: Routledge.

Hayes, M. (2006) Introduction: Teaching for Tomorrow, in M. Hayes and D. Whitebread (eds) *ICT in the Early Years*. Maidenhead: Open University Press.

Hoath, L. (2012) Learning Science Beyond the Classroom, in M. Dunne and A. Peacock (eds) *Primary Science: A Guide to Teaching Practice*. London: Sage.

Howard, D., A. Perry, M. Smith, L. Flintoft and R. Collins (2012) Linking Science to Numeracy and ICT, in M. Dunne and A. Peacock (eds) *Primary Science: A Guide to Teaching Practice*. London: Sage.

Johnston, J. (2005) *Early Explorations in Science* (2nd ed.). Maidenhead: Open University Press.

Johnston, J. (2011) Learning in the Early Years, in W. Harlen (ed.) *ASE Guide to Primary Science Education* (4th ed.). Hatfield: Association for Science Education.

Katz, L.G. (2010) How Young Children Learn, in S. Smidt (ed.) *Key Issues in Early Years Education*. Abingdon: Routledge.

Kennington, L. and J. Meaton (2009) Integrating ICT into the Early Years Curriculum, in H. Price (ed.) *The Really Useful Book of ICT in the Early Years*. London: Routledge.

Lowe, H. (2009) Children's Independence and ICT, in H. Price (ed.) *The Really Useful Book of ICT in the Early Years*. London: Routledge.

Miller, J. (1997) *Never Too Young: How Young Children Can Take Responsibility and Make Decisions. A Handbook for Early Years Workers*. London: National Early Years Network.

Ofsted (2009) *The Importance of ICT: Information and Communication Technology in Primary and Secondary Schools 2005/2008*. London: Ofsted Publications.

O'Hara, M. (2004) *ICT in the Early Years*. London: Continuum.

Porter, C. (2013) ICT Enhancing Science, in L. Kelly and D. Stead (eds) *Enhancing Primary Science*. Maidenhead: Open University Press.

Qualter, A. (2011) Using ICT in Teaching and Learning Science, in W. Harlen (ed.) *ASE Guide to Primary Science Education* (4th ed.). Hatfield: Association for Science Education.

QRStuff (2013) *QRStuff.com: Get Your Codes Out There!* (retrieved from www.qrstuff.com/qr_stuff.html).

Roden, J. (2011) Observation, Measurement and Classification, in H. Ward, J. Roden, C. Hewlett and J. Foreman (eds) *Teaching Science in the Primary Classroom: A Practical Guide* (2nd ed.). London: Sage.

Siraj-Blatchford, J. and D. Whitebread (2003) *Supporting Information and Communications Technology in the Early Years*. Buckingham: Open University Press.

UKCCIS (2012) *Advice on Child Internet Safety 1.0: Universal Guidelines for Provider* (retrieved from https://media.education.gov.uk/assets/files/ukccis%20advice%20on%20child%20internet%20safety.pdf).

Ward, H. (2011) The Use and Abuse of ICT, in H. Ward, J. Roden, C. Hewlett and J. Foreman (eds) *Teaching Science in the Primary Classroom: A Practical Guide* (2nd ed.). London: Sage.

Whitebread, D. (2008) Introduction: Young Children Learning and Early Years Teaching, in D. Whitebread and P. Coltman (eds) *Teaching and Learning in the Early Years* (3rd ed.). Abingdon: Routledge.

Whitebread, D., R. Dawkins, S. Bingham, A. Aguda and K. Hemming (2008) 'Our Classroom Is Like a Little Cosy House!': Organising the Early Tears Classroom to Encourage Independent Learning, in D. Whitebread and P. Coltman (eds) *Teaching and Learning in the Early Years* (3rd ed.). Abingdon: Routledge.

9
Planning early years science experiences
Di Stead

Introduction

The young learners at Rainhill Community Nursery were absorbed in their activities and only briefly looked up to see who the visitor was. There was a quiet purposeful buzz to the nursery. Some children were playing together in small groups, others played by themselves; some were involved in activities with an adult; some were inside, some were outside. It seems to fulfil Bryce-Clegg's (2012: 5) vision that 'Our ultimate aim is to inspire children to want to know more, to be eager to learn and be engaged in the opportunities that we create for them and, just as importantly, the ones we let them create for themselves.'

But how were the activities selected to do this? What were the principles underpinning what the children were doing? How did the setting plan to provide a curriculum which gives every child 'the best possible start in life and the support that enables them to fulfil their potential'? (DfE 2014: 1).

This chapter will consider *why* we plan and *how* we plan to develop young children as scientists and, throughout the chapter, a case study of the planning at Rainhill Community Nursery will help to illustrate some of these principles in practice.

Why plan? The rationale

It could be said that planning for science in the early years is unnecessary because children are naturally curious and born scientists and that science learning happens naturally as children are *constantly* making sense of the world they live in. Consequently, providing an interesting environment that motivates children is all that is required. If this really did develop children's thinking and ensured that effective use was made of children's time and key learning ideas were not left to chance then we would not need to plan, but we know this is not the case. For Rodger (2012: 39), good planning 'makes children's learning effective, exciting, varied and progressive, because it enables practitioners to build up knowledge about how individual children learn and make progress'. Woods (2013: 1 and 4) agrees that planning is central to the role of an early years practitioner because, to organize future learning and teaching which takes account of children's interests, careful consideration needs to be given to the physical learning environment, resources, learning outcomes, observation and assessment. For Bryce-Clegg (2012: 2) 'thoughtful planning ensures that children enjoy a variety of interesting experiences that will stretch their physical skills, social and communicative abilities and their knowledge of their own world.' Thoughtful planning of interesting science experiences which both enrich and challenge children's scientific thinking is a recurring theme in this book.

Principles which underpin planning early science experiences: what and how?

Planning for early science experiences, as for all learning, should consider how children might access the curriculum and reflect the characteristics of effective teaching and learning which are identified as:

- Playing and exploring: children investigate and experience things, and 'have a go.'
- Active learning: children concentrate and keep on trying if they encounter difficulties and enjoy achievements.
- Creating and thinking critically: children have and develop their own ideas, make links between ideas, and develop strategies for doing things. (DfE 2014: 1.9)

A number of common themes which guide planning for the early years can be seen in the different early years documents in the UK, which are that, when planning, adults must:

- consider and be responsive to the emerging needs, interests, talents and stage of development of each child
- reflect on the different ways that children learn from their experiences of the familiar world
- create rich, challenging and meaningful learning experiences
- encourage children to enjoy learning by asking questions and trying to find answers through exploring, enquiring, investigating and experimenting
- select activities which are challenging and provide an enjoyable experience both inside and out of doors in a safe and systematic way
- ensure a balance between activities led by children, and activities led or guided by adults.

It is important to bear in mind that we plan to facilitate children's learning. By putting children at the heart of the process, planning becomes an anticipation of learning rather than 'a pre-determined journey towards a goal that can be tracked, measured and accomplished' (Woods 2013: 1 and 4) where practitioners 'plan for endless possibilities' and have resources ready to follow the children's interests. She advocates that practitioners take a flexible and creative approach to planning which embraces the unexpected so that good use is made of 'magic moments' (Rodger 2012: 39). Planning for possibilities rather for outcomes means that rather than planning being 'an adult activity, where practitioners decide what should be learned, when and how' (Gripton 2013: 8) it becomes a process by which adults consider how they support and enable children's learning. However, although planning should arise out of the interests of the child, practitioners should also consider when it is the right time to extend children's experiences, and how to draw them into these new experiences (Rodger 2012: 39). She points out, as does Bryce-Clegg (2012: 2), that planning for new and interesting challenges is necessary because children's life experiences are limited. Consequently, planning should include what *we*, the more knowledgeable adults, want children to learn. The attitude of the adults to this flexible approach to planning is crucial because the children may not end up doing what was planned. Bryce-Clegg (2012: 4) reminds us that 'Every time you plan an activity on paper you only plan one possible outcome. When you then introduce your activity idea to children you should be prepared for a very distinct possibility that they will do something else with it.'

Key point

It is the children's learning which should be our main concern, not our plans, claims Hutchin (2000: 9). Brodie (2013) adds that our plans are there to support us, not to hinder us.

The importance of children's science learning being based in their own lived experiences has been highlighted throughout this book and is a feature not only of good practice within the early years but of effective science education (Harlen 2010: 10). When planning to extend children's knowledge and understanding of science ideas and help them work scientifically as they learn about their world, early years practitioners need to take account of:

- conceptual knowledge (an understanding of and about science)
- attitudinal knowledge (attitudes that underpin exploration and investigation)

- procedural knowledge
- the skills required to work scientifically.

(Brunton and Thornton 2010: 11)

As was mentioned in Chapter 1, the skills children use when playing, exploring and investigating are fundamental to science and are integral to learning science concepts. Consequently, opportunities for children to practise these skills should be considered when planning for early science experiences. These science skills include:

- Practical skills of observation: noticing, and looking closely and in a sustained way (sustained and focused observations helps children to *pay attention*).
- Being curious and asking questions.
- Reasoning and thinking skills: thinking about what will happen; explaining what they have seen.
- Noticing similarities and differences in order to sort and classify to support the development of ideas.
- Measuring: how high the bean stalk has grown, which car goes the furthest.
- Testing and finding out for oneself.
- Communication skills: speaking and listening, discussing, describing and recording observations, using correct vocabulary.
- Social skills of cooperation, negotiation, leadership, following instructions.
- Behaving in a safe manner.

(after Brunton and Thornton 2010: 15 and Glauert et al. 2007: 132)

Creating opportunities to extend children's knowledge and understanding of science ideas requires thoughtful planning on the part of early years practitioners and is informed by their own knowledge of the science concept. As has been highlighted throughout this book, practitioners need to plan a rich learning environment which offers breadth as well as depth (Rodger 2012: 181) and include a variety of activities that reinforce the science idea which is the focus of children's learning. Using this approach, practitioners plan for possible lines of development (PLODs) and have in mind questions which challenge children's thinking. Examples of such questions are given by Faith Fletcher in Chapter 5 and Babs Anderson in Chapter 3. Thinking at the planning stage of the range of activities which both engage and challenge enables early years practitioners to take a flexible approach when supporting children's learning.

Working with associate teachers has also highlighted for me the importance of planning to foster positive attitudes to science. These attitudes, which are developed through explorative play and practical experience, are:

- Wanting to find out an answer to a question.

- Being excited about what will be found out.

- Feeling awe and wonder.

- Persevering.

- Having a caring attitude to living things.

Preparing to plan for early science experiences?

Based on my experience of working with associate teachers preparing to work with young children, there is a view that developing knowledge and understanding of science ideas in young children is not a priority and that it is more important to focus on providing experiences that focus on exploring and investigating (see Chapter 1). However, as has been highlighted throughout this book, children are constantly developing their knowledge and understanding of science ideas through their lived experience. While acknowledging that 'no one can be an expert in all areas' (Stone 2012: 3) it does not seem unreasonable that significant adults should know the background science for the areas of science relevant to children in the early years and how selected activities may help children develop key scientific ideas that lay secure foundations for later science learning. For example, why do we provide *gloop* for children to explore? Why do we give children shiny and dull materials to examine or why do we say that some colours are bright? Having secure science background knowledge also gives early years practitioners confidence to consider how ideas and activities taken from published frameworks might meet the wider aims of science education or the children's individual learning needs and so overcome concerns raised by Ofsted (2013: 23). Developing secure science subject knowledge for the areas of science that are relevant to the early years also enables practitioners to plan for PLODs, which ensures the breadth and depth of science experiences these children need, as discussed in Chapter 1. So how do early years practitioners both check their background knowledge and ensure that the activities they choose meet the needs of the children in their setting? The Internet provides easily accessible information for teachers; however, care needs to be taken to ensure that the source is reliable and the science information is correct. Using background knowledge provided for early years practitioners, rather than relying on the tips to pass GCSE science, will help practitioners choose age-appropriate examples.

Another factor to be considered is whether to plan for areas of learning, specifically in this instance early science experiences, separately or whether to take a more holistic approach. Rodger (2012: 181) acknowledges that labelling subjects helps to inform assessment and planning but argues that children do not make this distinction between subjects or areas of learning. According to Tina Bruce (1997) subjects cannot be separated because skills and understanding are interlinked. Max de Bóo (2004) agrees and says that children don't see the difference between 'work' and 'play' or between language development and scientific development. Kelly and Stead (2013: 5) agree that children gain a holistic view of learning when subjects are taught in an integrated way. However, they argue that being aware of the distinctive features of science when planning is essential, when science is being taught in a cross-curricular way, to ensure that subject specific skills are developed and science maintains its identity.

Associate teachers often raise concerns about risks and keeping safe when discussing planning for early science experiences. It has already been acknowledged that children need to take risks (Chapters 4 and 5) and /'science provides an opportunity for children to develop an understanding of safe ways of working and of using equipment. Children need to know when to stop and think about safe practices. Bianchi and Feasey (2011) advocate involving children in the risk assessment, saying this offers children the chance to take responsibility and consider how to manage risk not only to themselves but also to others they are working with. Their 'Children's Risk Management Card' (2011: 65) can be used as a prompt to aid discussion with children. Children are asked to identify any hazards (anything that can cause harm), to say why it may cause harm and finally to suggest ways they can keep safe. This discussion can also help children feel confident about communicating their concerns to others.

It is a teacher's responsibility to ensure that they have considered any risk presented by the activities they have planned for children. Although I recognize that there is a general concern among the public about safety in science, the early years setting is unlikely to be using dangerous substances or asking children to perform dangerous activities. We need to be careful that we do not succumb to health and safety myths and use safety as an excuse to avoid practical experience. The Association for Science Education's *Be Safe!* (ASE 2011) provides advice and guidance to ensure that the risks are minimized.

Good practice in planning

Planning is a process that involves thinking, discussing, doing and reflecting. Young children become part of this process, showing you their interests and preferences, by their actions just as much as their words, when spoken language develops. Adult planning energy will have created an accessible, well resourced learning environment, indoors and outdoors.

(Bryce-Clegg 2012: 3)

Planning works best when all adults in the setting are involved with the process because it provides an opportunity for the whole team to clarify their thinking and develop a shared understanding about the philosophy and approach which under-pins their practice (Rodger 2012; Woods 2013). This is central to the planning carried out at Rainhill Community Nursery. It is common place for longer term plans to arise out of the statutory documents because teachers have a duty of care to ensure any statutory requirements are met (Bryce-Clegg 2012: 6). Weekly plans then build on observations and judgements, and planning is 'translated into practice which is then evaluated according to whether the child has met the learning goals set' (Gripton 2013: 8). However, given that it is considered good practice to follow children's interests, how do we do that?

An essential part of planning is gathering information which will enable practitioners to plan to 'excite children, encourage exploration and play, challenge and foster concentration and encourage new ideas, applying what they already know' (Rodger 2012: 38). When planning for early science experiences it is also helpful be aware of the science ideas that research tells us young children may have developed (see the SPACE project, www.nuffieldfoundation.org/primary-science-and-space). Planning and assessment working 'hand in glove' (Rodger 2012: 39) is seen as good practice and is highlighted in the case study at Rainhill Community Nursery. The headteacher emphasized that 'meticulous tracking is essential even before planning can begin'. Importantly, tracking in this setting is not seen as a separate entity, but part of the planning and assessment cycle. It focuses on what the children can do – in contrast to a deficit model, which looks at what they cannot do. It also helps staff find out what areas are not being addressed. Observations of the children are shared by the staff at Rainhill Community Nursery to build up a rounded picture of what science the children are learning. Discussions explore what the evidence has told them so they can plan what needs to happen next. The headteacher emphasized that it was the shared judgements, involving all the staff, which were central in deciding what to do next.

Task 9.1

Finding out what children know and are interested in

- How do *you* find out what children can do or know in science?
- Can you think of a time when *you* made a judgement, from an observation, about what interested a child?
- Do you need a different technique to find out what children know and can do in science compared to other areas of learning?

Planning should ensure that activities are matched to their purpose – to support a small group, a class group or an individual, whether in child-sustaining or adult-led activities – and consider the role of the adults. Brunton and Thornton (2010: 17) suggest that the role of the adult is to:

- provide interesting starting points
- see potential for child-initiated play
- probe children's thinking.

The point at which the adult should intervene to probe understanding (see Chapter 5) should be considered at the planning stage. This planning for supporting adults should consider what questions to ask, what explanations to give, what teaching points need making or what evidence needs to be collected.

Key point

Rodger (2012: 38) says that there should be a balance between what children choose to do themselves and what adults generally regard as worthwhile, planned activities that require adult support or timely intervention.

Evaluation

Evaluation is part of the planning cycle and is informed by analysing the assessments and observations of children's learning. Questions which guide this analysis are:

- What does this tell me about the child's learning and development?
- How did the children respond?
- Did it meet their needs and interests?
- What shall we offer next?

(Hutchin 2013: 125)

Babs Anderson in Chapter 3 also offers advice for evaluating planning.

Task 9.2

Choose a piece of medium-term planning that you have used or are about to use.

Highlight the parts of the plan which have a science focus. Look at the planning and see if you can identify these 'top tips'. Does your planning:

- include talk with parents to find out what children are interested in and know?
- ensure all children can access the activities?
- tune in and following children's interest?
- build confidence and a 'can-do' attitude to solving science problems and answering science questions?
- encourage the children to make their own choices?
- provide a stimulating environment for learning science?
- include starting points which motivate and challenges children's thinking?
- allow time for sustained thinking?
- remember that play and exploration are fundamental to learning and thinking?

After Hutchin's 'Top Tips for Effective Practice'
(2013: 18–19)

How does the planning provide the children with the opportunity to:

- Explore?
- Play?
- Follow their interests?
- Find answers to questions?
- Practise science skills?
- Talk about what they have seen?
- Make sense of what they see?
- Use ICT?

After Hutchin's table 1.1, 'Characteristics of
Effective Learning' (2013: 9)

> **Key points**
>
> - Be brave enough to step away from the written plan. If you don't do what is in your plan, don't worry (Hutchin 2000; Woods 2013: 2).
> - It is the process of planning not the outcome that is important.
> - Before you plan, consider what you know about the children: what do they already know, what can they do and what they are interested in?
> - Consider all *possibilities* in your planning.

9.1 Case study

'Open the door and let the children come through'

'Learning is paramount at Rainhill Community Nursery', said the headteacher 'and the children making progress is a prime concern.' I was in the nursery to find out how one nursery set about planning for science learning in their curriculum. The headteacher said that she thought that there was not one way to plan. 'It's all about children learning', she said; 'they are the most important factor in the planning.' The view that early years practitioners must have high expectations underpins her approach, and is supported by Thornton and Brunton (2013). She has similar ambitions to the headteacher at Hursthead Junior School (Kelly and Stead 2013: 139), who said, 'Staff need to have the courage to have high expectations and not to limit their expectations, for example because of the background of the children, and to be brave enough to not only reach, but to exceed expectations.'

The headteacher at Rainhill Nursery thinks that using the child's interest makes the job of planning easier. She says that 'A child will enjoy and learn better if they are interested. Children have a basic entitlement to learn and we are guided by early learning goals. Children need permission to do real things; things which fascinate them, things they cannot do at home.' Laughing, she says there is at least one occasion each year of a child cutting another child's hair. She adds, 'We have to be pragmatic; following a child's interest does not mean that you have to follow 80 interests.' You may not be able to use every current interest, but the staff 'just do our best'. She elaborates, 'The main pitfall is kidding ourselves that we are following the child's interest when we are not. You must ask whose interest we are following. The activities are meant to serve the children, not the adults.'

Task 9.3

Outside one early morning in summer, one child pointed out the morning chorus. The adult mused 'I wonder how many birds there are.' The child suggested that they counted the tweets. Five tweets were heard and the children suggested that there must be five birds singing. It would have been easy to become bogged down in a discussion about whether one tweet equated to one bird, but instead the adult listened to further questions the children raised. 'Why are they hungry?', 'What do they eat?', 'Worms?' The children had been stimulated by the birdsong that morning and had plenty to talk about.

- How the adult uses this starting point to plan what happens next is crucial. What would you plan to do next?

Context of Rainhill Nursery

Eighty children attend Rainhill Nursery. It is staffed by four teachers with qualified teacher status (the headteacher and two part-time and one full-time teachers) and seven support staff who are qualified to at least Level 2 (NVQ or equivalent), but mainly Level 3.

The headteacher described the catchment area as 'a whole world in a village'. She added that 'Experience shows us that the area the nursery serves tends to be populated with a wide range of growing families, people who have brought up their families and are new grandparents, and commuters to Liverpool and Manchester. Often one or both of the children's parents are working. However, demographic information shows us there is a real range of people in the neighbourhood: some looking for work and some who are quite affluent. Considering the size of the village and the demographic makeup of St Helens, a wider range of ethnic backgrounds is represented here than you would expect. Each year there is a similar percentage of children with English as an additional language [EAL], predominantly from Asia and the Middle East: Urdu speakers, Chinese and Tamil.'

Background of headteacher

The headteacher has been in at Rainhill Nursery for eight years. She had previously taught in a primary school at a children's centre in an economically deprived area in Liverpool, where the children 'had bags of potential'. When she first arrived at the children's centre she was shocked by the view of some members of staff, that expectations would have to be lowered because of the area from which the children were coming. This reinforced her belief that the aim of the provision in a nursery setting was to 'open the door and let the children come through'. This philosophy drives the headteacher's approach to planning, but she acknowledges that the team are the ones who put the policy into practice. She said that the children have an 'entitlement to be inspired'.

The headteacher has a degree in mathematics and although she does not draw often on these studies in her early years work, she believes that the greater depth of understanding has given her comfort and confidence in planning for children's mathematical development. Therefore, she believes, further scientific study and background knowledge are bound to be of benefit to those working with young children.

The curriculum: 'entitlement to be inspired'

The curriculum provided in many ways follows what is outlined in the earlier part of this chapter.

Rainhill Nursery aims to provide a creative curriculum. The context for learning is decided by the staff and the children, and is ambitious. Planning:

- *is concerned with where the child is now*
- *is concerned with what the child needs to learn next*
- *follows the child's interests*
- *also includes what teachers think children need to know.*

Although science is not taught discretely here the learning environment fosters positive attitudes to learning in science. First-hand experience is valued because it allows children to be inquisitive and helps them explore their world. Planning staff are mindful of the child's natural desire to explore in a sensory way and know how this might help a child cognitively. Activities are selected because they enable children to learn science skills such as being curious and observing closely which will equip them to learn in the future. The specific science skill of enquiry is central; children are encouraged to suggest what might happen and finding out answers for themselves by testing ideas. Language also plays a crucial role in children's learning of science ideas, and the use of specific vocabulary is important, as pointed out by Babs Anderson in Chapter 3.

How do the staff at Rainhill Nursery make this happen?

Planning meetings, which all staff are contracted to attend, take place fortnightly with a review meeting at the end of the first week, which as many staff as possible are expected to attend.

Planning here starts with what the child needs to learn. The staff share their observations of the children to build up a rounded picture of what science the children are learning. A discussion explores what the evidence has told them and to plan what needs to happen next. Collecting information about what the adults know about their children is necessary to enable staff not only to support children and provide what they need, but also to initiate learning. This can provide the children with the opportunity to apply what they have learnt. Parents have a role in planning, too. Informally, parents tell staff about what interests their child has and what the child talks about at home. Making time for assessment and moderation to check judgements across all adults are valid is prioritized because planning builds on this information.

The benefits of planning in this way are that the professional knowledge of a diverse and experienced team is maximized. Consequently, the needs of the child, and their interests, are more likely to be addressed. Differentiation is a big feature of the planning and depends on accurate assessment.

The headteacher says that observation is key here. 'Objectives need to be qualitative, to tell what you actually saw the child do, or heard them say, rather than be a "judgement" – that's different!' An example of this is given by Faith Fletcher in Chapter 5. The assessment that is carried out uses the observation as evidence to support the judgement. 'We want to show what children can do, rather than adopting a deficit model, which is can't do this, can't do that. The observations also enable us to make sure that we incorporate the child's voice, because we can see and hear what makes them tick!'

Short-term planning is displayed on a board (Figure 9.1) and is available to a number of different audiences, including all staff and parents. This is seen as good practice by Kathy Brodie (2013: 86). The planning clearly outlines what each group will do.

Staff annotate the planning on the display board as soon as they notice a child's response (Figure 9.2). Staff should not act in isolation but draw on what parents say and what any visitors may notice. Tracking also helps staff find out what areas are not being addressed.

Science subject knowledge of the adults is recognized as important because practitioner confidence and competence are crucial. Adults need to feel comfortable about the science they are teaching. They need to know how both the scientific ideas and the science skills being developed in the selected activities will be useful in the future. The strongest practitioners not only have this confidence and competence but also have an accurate knowledge of child development and where the children are in their learning. The enthusiasm of staff can inspire children and their interests can be seen as a strength. For example, an adult's love of gardening or cooking helps to authenticate the experience and bring a real-life perspective.

The headteacher elaborate and asks how much the resources represent real life. How far do they help children understand everyday science? For example, role play can represent real life and may help children apply science in everyday situations but in some cases a science role-play area may just turn out to be a place where buttons are pushed. It may not help children understand either a science idea or the role of the scientist. She laughs at the times they built a space rocket and the children made it into a tea room. I wonder if this is because,

Figure 9.1 Planning in the nursery is displayed for all to see.

unlike the children Jessica Baines- Holmes describes in Chapter 6, these children were not involved on the planning and construction of the space rocket.

The setting has started to implement the Nature Kindergarten philosophy from MindStretcher introduced by Claire Warden (www.mindstretchers.co.uk/ nature-kindergartens.cfm). There is an emphasis on the environment and the children's awareness of sustainability and children are encouraged to explore,

Week beginning: 7th October

Area B: Investigative area, maths area, construction are, computers, writing table, observational table, book corner, role-play area.

Intended learning outcomes	Enhancement to provision	Adult-focused activities and experiences	Evaluation/observations
Objectives To talk about some things they have observed such as plants, animals and natural and found objects (UW: The World pg 41) To question why things happen and to give simple explanations (C and L: Speaking pg 21) <u>Less confident children</u> To explore the resources present <u>More confident children</u> To look closely at similarities, differences, patterns and change <u>Lewis</u> To sometimes make enquiries on particular features/processes <u>Children with EAL</u> To use some language to describe objects	**Investigative table** Autumn display: Leaves Conkers Acorns Photographs Fiction ad non-fiction books Autumn jigsaw Clipboard and pencils Magnifying lenses **Computer** Digital viewer set up on computer with basket containing autumnal objects **Observational table** Giant African land snails	Adults to bring the children's attention to the various resources on the autumn display. Pick up resources modelling exploring them using the senses, e.g. rustling leaves near the ear. Adults model appropriate language to describe the objects and to encourage the children to talk about the object they are exploring Model the books. Bring children's attention to front cover and the index page placing your finger under the text as you read Encourage children to explore the autumnal resources using digital viewer. Model use of digital viewer and show how to capture image of the objects. Encourage children to talk about their findings. Adult to model appropriate descriptive language	There has been a lot of interest in the autumn display this week. Many children used the magnifying lenses to further explore the features of objects. Children have been involved with exploring red paint this week. Now add red leaves to the investigative table.

Figure 9.2 Annotated planning at Rainhill Community Nursery.

appreciate and respect their environment. One technique used is for children to look through a hoop or a pipe at the world to help focus attention and encourage closer examination of flower heads or rocks, for example (Figure 9.3). The children might look at sheep's wool, felting and teasing the wool, for example, or the features of a leaf might be highlighted using a light box.

Last year the nursery introduced floor books to capture the children's thinking: a collaborative venture with both children and adults contributing and

Figure 9.3 Looking down a drainpipe helps a child to focus their observation and pay attention.

with science providing the context. Children are enthusiastic co-authors of these
books and use them regularly to review and revisit their learning (Figure 9.4).

Floor books can be used beyond nursery. An example of this can be seen in
Figure 9.5 at Anfield Infant and Early Years School.

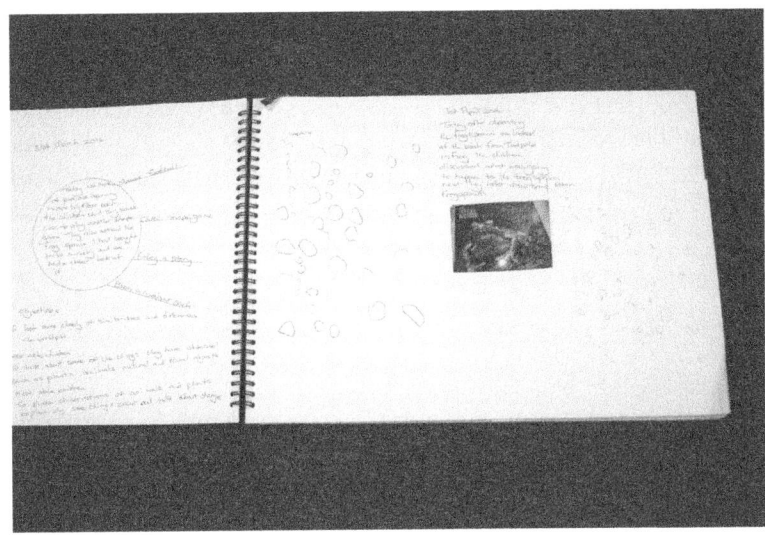

Figure 9.4 Floor book from Rainhill Community Nursery.

Figure 9.5 Floor book from Anfield Infant and Early Years School.

Conclusion

Planning tell us a great deal about the values and attitudes of a setting. Working together as a team, as demonstrated in the case study of Rainhill Community Nursery, helps put the philosophy into practice. Adults can contribute their own expertise and interests. A planning meeting provides an opportunity to share knowledge of the children and moderate judgements. Planning should ensure that the children are provided with new and challenging science experiences from which to develop scientific ideas, practise skills of exploration and foster a positive attitude to science. Planning can also provide children with opportunities and inspiration to take control of their own learning by following their interest.

References

ASE (Association for Science Education) (2011) *Be Safe!* (4th ed.). Hatfield: ASE.

Bianchi, L. and R. Feasey (2011) *Science Beyond the Classroom Boundaries for 3–7 Year Olds*. Maidenhead: Open University Press.

Brodie, K. (2013) *Observation, Assessment and Planning in the Early Years: Bringing It All Together*. Maidenhead: Open University Press.

Bruce, T. (1997) *Early Childhood Education*. London: Hodder & Stoughton.

Brunton, P. and L. Thornton (2010) *Science in the Early Years*. London: Sage.

Bryce-Clegg, A. (2012) *Planning for Early Years: Garden and Growing*. London: Practical Pre-School Books.

de Bóo, M. (2004) *The Early Years Handbook*. Sheffield: Curriculum Partnership.

DfE (Department for Education) (2014) *Statutory Framework for Early Years and Foundation Stage: Setting the Standards for Learning Development and Care for Children from Birth to Five*. London: DfE (retrieved from www.gov.uk/government /publications/early-years-foundation-stage-framework—2).

Glauert, E., C. Heal and J. Cook (2007) Knowedge and Understanding of the World, in J. Riley (ed.) *Learning in the Early Years: A Guide for Teachers of Children 3–7* (2nd ed.). London: Sage.

Gripton, C. (2013) Planning for Endless Possibilities, in A. Woods (ed.) *Child-Initiated Play and Learning*. London: Routledge.

Harlen, W. (ed.) (2010) *Principles and Big Ideas of Science Education*. Hatfield: ASE (retrieved from www.ase.org.uk).

Hutchin, V. (2000) *Tracking Significant Achievement in the Early Years* (2nd ed.). London: Hodder Murray.

Hutchin, V. (2013) *Effective Practice in the Early Years Foundation Stage*. Maidenhead: Open University Press.

Kelly, L. and D. Stead (2013) *Enhancing Primary Science: Developing Effective Cross-Curricular Links*. Maidenhead: Open University Press.

Ofsted (2013) *Maintaining Curiosity in Science*. Manchester: DfE (retrieved from www. ofsted.gov.uk/resources/130135).

Rodger, R. (2012) *Planning an Appropriate Curriculum in the Early Years: A Guide for Early Years Practitioners and Leaders, Students and Parents*. London: David Fulton.

Stone, T. (2012) *Planning for Early Years: Homes and Families*. London: Practical Pre-School Books.

Thornton, L. and P. Brunton (2013) *16 Steps to Achieving an Outstanding Early Years Ofsted Judgement*. London: Optimus Education (retrieved from www.optimus-education.com/16-steps-achieving-outstanding-early-years-ofsted-judgement).

Woods, A. (ed.) (2013) *Child-Initiated Play and Learning Planning for Possibilities in the Early Years*. London: Routledge.

Index

Page number of figures and tables are included in **bold**

CHARACTERISTICS OF EFFECTIVE EARLY LEARNING
Helping young children become learners for life

Helen Moylett

9780335263264 (Paperback)
2013

eBook also available

The key argument of *The Characteristics of Effective Early Learning* is that how children learn is as important as what they learn. This book helps you understand how to support the learning and development of young children through promoting the characteristics of effective early learning: play and exploring, active learning, and creating and thinking critically.

Key features:

- Investigates how children engage in learning through playing and exploring, and are motivated through active learning
- Explores how children become creative and critical thinkers able to review their own learning and thinking, imaginatively solving problems and excited by their own
- Examines appropriate approaches to observation, assessment and planning

www.openup.co.uk

 Open University Press
McGraw - Hill Education